I0004095

System Design Guide for Software Professionals

Build scalable solutions – from fundamental concepts to cracking top tech company interviews

Dhirendra Sinha

Tejas Chopra

System Design Guide for Software Professionals

Copyright © 2024 Packt Publishing

All rights reserved. No part of this book may be reproduced, stored in a retrieval system, or transmitted in any form or by any means, without the prior written permission of the publisher, except in the case of brief quotations embedded in critical articles or reviews.

Every effort has been made in the preparation of this book to ensure the accuracy of the information presented. However, the information contained in this book is sold without warranty, either express or implied. Neither the authors, nor Packt Publishing or its dealers and distributors, will be held liable for any damages caused or alleged to have been caused directly or indirectly by this book.

Packt Publishing has endeavored to provide trademark information about all of the companies and products mentioned in this book by the appropriate use of capitals. However, Packt Publishing cannot guarantee the accuracy of this information.

Associate Group Product Manager: Kunal Sawant
Publishing Product Manager: Akash Sharma
Book Project Manager: Deeksha Thakkar
Senior Editor: Esha Banerjee
Technical Editor: Vidhisha Patidar
Copy Editor: Safis Editing
Proofreader: Esha Banerjee
Indexer: Hemangini Bari
Production Designers: Joshua Misquitta and Ponraj Dhandapani
DevRel Marketing Coordinator: Sonia Chauhan

First published: August 2024

Production reference: 1300724

Published by Packt Publishing Ltd.
Grosvenor House
11 St Paul's Square
Birmingham
B3 1RB, UK

ISBN 978-1-80512-499-3

www.packtpub.com

To my parents, whose unwavering support and love have been my greatest source of strength. Your belief in me has always been the foundation upon which I've built my dreams.

To my lovely wife, Shruti, for her endless patience, understanding, and constant encouragement. Your love has been my guiding light, and your unwavering support has given me the strength to persevere.

To my kids Kunshi and Kavni, for their boundless energy and joy, which have been a constant source of inspiration. Your curiosity and enthusiasm for learning inspire me every day.

To my teachers, who imparted wisdom and inspired a lifelong love of learning. Your dedication and passion for teaching have left an indelible mark on my life and career.

To all the book authors, blog writers, and content creators who have expressed their thoughts and shared their expertise in the area of software system design. Your insights and contributions have been invaluable, and this book is a testament to the collective knowledge and creativity of our community.

Dhirendra Sinha

To Mahima, my loving wife and steadfast partner in this journey – your unwavering support and encouragement have been my anchor. And to our little wonder, Nyra – your curiosity lights up my world and fuels my passion to explain even the most complex ideas.

To my parents and sister, Anchal – your love, guidance, and belief in me have shaped not just this book, but the person I am. Your support has been the wind beneath my wings, pushing me to explore new heights in technology and life.

This book is a labor of love, inspired by and dedicated to all of you. Thank you for being my strength, my motivation, and my joy.

Tejas Chopra

Contributors

About the authors

Dhirendra Sinha is a Software Engineering Manager at Google. He is an angel investor and has served as a Strategic and Technology advisor at a few startups. Dhirendra has around two decades of experience in the Software Engineering field building highly scalable complex distributed systems and managing multiple Engineering teams. While he has experience in working at large companies such as Cisco, Oracle, Yahoo, and Google, he has also been involved with early and late stage startups in leadership positions. In addition to his strong Software Engineering experience, he has been teaching the Distributed System Design course for the last seven years. He has also been coaching and mentoring Software Engineers and Software Engineering Managers for over a decade. He completed his Bachelor of Technology degree from IIT Guwahati and Master of Science from Texas A and M University, College Station, Texas.

Tejas Chopra is a Senior Engineer at Netflix working on building the Machine Learning Platform powering Netflix recommendations and personalization. He is also a Co-founder at GoEB1, which is the world's first thought leadership platform for immigrants. Tejas is a recipient of the prestigious EB1A (Einstein) visa in the US. He is a Tech 40 under 40 Award winner, a 2xTEDx speaker, a British Computer Society (BCS) Fellow, and has spoken at conferences and panels on Cloud Computing, Blockchain, Machine Learning, Software Development, and Engineering Leadership. Tejas has been awarded the "International Achievers Award, 2023" by the Indian Achievers' Forum. He is an Adjunct Professor of Software Development at the University of Advancing Technology, Arizona, an Angel Investor, and has previously been an Advisor to startups such as Nillion, Inc. He is also a member of the Advisory Board for Future of Memory and Storage Summit. Tejas' experience has been in companies such as Box, Apple, Samsung, Cadence, and Datrium. Tejas holds a Masters Degree in Electrical & Computer Engineering from Carnegie Mellon University, Pittsburgh.

About the reviewers

Ruchi Agarwal is a Senior Software Engineer currently working as a technical lead at Netflix, with over 13 years of experience in building and leading technical projects across multiple high-tech companies, including Netflix, Apple, and eBay

At Netflix, Ruchi was the first hired engineer in the internal employees' applications team that builds solutions for company-wide usages. Building the team from its inception, she played a critical role in setting the engineering direction and building essential tools to streamline Netflix's corporate operations for its employees.

As a technical leader, Ruchi has spearheaded numerous projects integral to the operations of Netflix, which employs over 12,000 people. Ruchi led and architected numerous projects at Netflix that saved the company millions of dollars in annual revenue.

During her time at Apple, Ruchi was a critical resource in building tools that played a crucial role in the development of new iPhones.

Ruchi actively champions the empowerment of women in the tech industry and inspires students through impactful speaking engagements ensuring that the voices and perspectives of women continue to shape the future of technology

Chinmay Abhay Nerurkar is a highly respected Engineering Leader with 15+ years of diverse industry experience in building enterprise-scale software products using AI/ML, Big Data, Distributed System Design, and Cloud technologies. He is currently a Principal Engineer at Microsoft AI, leading strategic engineering initiatives on the Microsoft Copilot Platform.

Chinmay is an experienced career mentor who coaches engineering managers and individual contributors on all-rounded growth. His passion lies in cultivating an inclusive, empowering team culture while promoting engagement and driving innovation. Chinmay is also a sought-after keynote speaker who has presented at major international conferences such as P99 CONF, DeveloperWeek NYC, AI DevWorld SFO, Graph+AI ML, etc. and has chaired multiple industry advisory boards. He is a Fellow at the Institution of Engineering and Technology and at the British Computing Society, Senior Member at IEEE, and Senior Fellow at AI 2030 think-tank for Responsible AI.

Table of Contents

Part 2: Core Components of Distributed Systems

4

Distributed Systems Building Blocks: DNS, Load Balancers, and Application Gateways   53

5

Design and Implementation of System Components –Databases and Storage   73

6

Distributed Cache   113

7

Pub/Sub and Distributed Queues   131

Part 3: System Design in Practice

8

Design and Implementation of System Components: API, Security, and Metrics   149

9

10

11

12

Designing a Service Like Instagram   227

13

Designing a Service Like Google Docs   253

14

Designing a Service Like Netflix   289

15

Tips for Interviewees   323

16

System Design Cheat Sheet   331

Index   343

Preface

Hello there! System Design is a crucial skill in the world of software engineering and architecture. It involves the process of defining the architecture, components, modules, interfaces, and data for a system to satisfy specified requirements. This book aims to provide a comprehensive guide to system design, covering both theoretical concepts and practical applications through real-world examples.

The field of system design is vast and ever-evolving, with new technologies and paradigms emerging constantly. As software systems become increasingly complex and distributed, the ability to design scalable, reliable, and efficient systems has become extremely important. This book strives to equip you with the necessary knowledge and tools to tackle complex system design challenges in today's technology landscape.

There are three main aspects that this book focuses on:

- Fundamental concepts and principles of system design

- Key components and technologies used in modern distributed systems

- Real-world system design case studies and their analysis

We will cover these aspects in-depth, providing you with a solid foundation in system design theory while also offering practical insights drawn from industry experience and best practices.

Throughout this book, we will provide relevant information and guide you through various system design concepts, techniques, and case studies. The content is based on two main sources of information:

- Our experience and knowledge from years of working in the software industry and designing large-scale systems

- Interviews and insights from industry experts and system design practitioners across different domains and companies

As per recent industry reports, the demand for professionals skilled in system design has been growing rapidly. Companies of all sizes, from startups to tech giants, are constantly seeking engineers who can architect and design scalable, robust systems. This trend is expected to continue in the coming years, as businesses increasingly rely on complex software systems to drive their operations and innovations.

One of the major challenges faced by many aspiring system designers is bridging the gap between theoretical knowledge and practical application. This book aims to address this challenge by not only explaining the core concepts but also demonstrating how these concepts are applied in real-world scenarios. By the end of this book, you should be well-equipped to tackle system design interviews and real-world design problems with confidence.

Who this book is for

This book is designed for a wide range of readers in the software engineering field. The three main personas who are the target audience of this content are as follows:

- **Software engineers and developers**: Those looking to expand their skills beyond coding and keen on understanding how large-scale systems are designed and architected. This book will help them grow in their career toward becoming senior engineers, tech leads, and system architects.

- **System design interview candidates**: Professionals preparing for system design interviews at top tech companies. The book covers common interview topics and provides a structured approach to solving design problems.

- **Engineering managers and tech leads**: Those who want to gain a deeper understanding of system design principles to better guide their teams and make informed architectural decisions.

- **Computer science students**: Advanced students who want to supplement their theoretical knowledge with practical insights into how real-world systems are built.

What this book covers

Chapter 1, Basics of System Design, provides an introduction to the field of system design. It explains the different types of system designs and emphasizes the importance of system design in the industry. This chapter sets the foundation for the rest of the book.

Chapter 2, Distributed System Attributes, delves into the fundamental concepts that underpin modern system design. It covers crucial topics such as consistency, availability, partition tolerance, latency, durability, reliability, and fault tolerance. Understanding these principles is essential for designing robust and scalable systems.

Chapter 3, Distributed Systems Theorems and Data Structures, explores the theoretical underpinnings of distributed systems. It covers important theorems and algorithms such as the CAP theorem, PACELC theorem, Paxos and Raft algorithms, and the Byzantine Generals Problem. It also introduces key concepts such as consistent hashing, Bloom filters, and HyperLogLog, which are frequently used in large-scale system design.

Chapter 4, Distributed Systems Building Blocks: DNS, Load Balancers, and Application Gateways, focuses on the core components of networked systems. It provides an in-depth look at DNS, load balancers, and application gateways, which are crucial for building scalable and reliable distributed systems.

Chapter 5, Design and Implementation of System Components – Databases and Storage, is dedicated to the various types of databases used in modern system design. It covers relational and NoSQL databases and dives into specific technologies such as Cassandra, HBase, DynamoDB, and S3. It also includes a section on designing a key-value store and an overview of Lucene search.

Chapter 6, Distributed Cache, explores the world of distributed caching. It covers the design of distributed cache systems and provides detailed information on popular caching solutions such as Redis and Memcached.

Chapter 7, Pub/Sub and Distributed Queues, focuses on designing distributed queues and publish-subscribe systems. It provides an in-depth look at technologies such as Kafka and Kinesis, which are crucial for building real-time data processing systems.

Chapter 8, Design and Implementation of System Components: API, Security, and Metrics, covers the essential aspects of designing and maintaining APIs in distributed systems. It explores REST and gRPC protocols, API security basics, and the crucial components of system observability: logging, metrics, alerting, and tracing.

Chapters 9 to 16 are dedicated to real-world system design case studies. Each of these chapters follows a consistent structure:

- Requirements of the system
- High-level design
- Detailed design
- Evaluation of the design

These case studies cover a wide range of popular systems:

- *Chapter 9: System Design – URL Shortener*
- *Chapter 10: System Design – Proximity Service*
- *Chapter 11: Designing a Service Like Twitter*
- *Chapter 12: Designing a Service Like Instagram*
- *Chapter 13: Designing a Service Like Google Docs*
- *Chapter 14: Designing a Service Like Netflix*
- *Chapter 15: Tips for Interviewees*
- *Chapter 16: System Design Cheat Sheet*

> **Note**
>
> By working through these case studies, readers will gain practical experience in applying system design principles to real-world scenarios. These chapters will help bridge the gap between theory and practice, providing invaluable insights into how large-scale systems are designed and implemented in industry.

The book concludes with guidance on approaching system design interviews. This section covers the format of system design interviews, what interviewers are looking for, and the importance of this type of interview in the hiring process. It also provides tips on asking relevant questions, considering boundary conditions, making back-of-the-envelope calculations and estimations, and designing systems based on access patterns.

To get the most out of this book

To fully benefit from this book, readers should have a basic understanding of computer science concepts, data structures, and algorithms. Familiarity with at least one programming language and basic networking concepts will also be helpful. However, we've strived to make the content accessible to readers from various backgrounds by explaining concepts clearly and providing the necessary context.

We recommend reading this book sequentially, as the later chapters build upon concepts introduced in the earlier ones. However, experienced readers may choose to jump directly to specific topics or case studies of interest. As you progress through the book, we encourage you to actively engage with the material. Try to solve the design problems presented in the case studies before reading the proposed solutions. This approach will help you develop your own system design thinking and problem-solving skills.

Remember that system design is as much an art as it is a science. While this book provides a solid foundation and numerous examples, there's often no single "correct" solution to a design problem. The best designs often emerge from balancing various trade-offs and considering the specific context and requirements of each situation.

We've also included numerous diagrams and illustrations throughout the book to help visualize complex concepts and system architectures. These visual aids are crucial in system design, both for understanding and communicating ideas effectively.

Finally, system design is a field that's constantly evolving. While this book covers the fundamental principles and current best practices, we encourage you to stay curious and continue learning beyond this book. Keep up with industry trends, new technologies, and case studies of how leading companies are solving their scaling challenges.

We hope this book serves as a valuable resource in your journey to mastering system design. Whether you're preparing for interviews, looking to advance in your career, or you are simply passionate about building large-scale systems, we believe you'll find the content both informative and practical.

Thank you for choosing this book. We're excited to embark on this learning journey with you. Let's dive in and explore the fascinating world of system design together!

Conventions used

There are a number of text conventions used throughout this book.

`Code in text`: Indicates code words in text, database table names, folder names, filenames, file extensions, pathnames, dummy URLs, user input, and Twitter handles. Here is an example: "The `Comment` entity represents the comments posted on photos. It contains the comment text, `user_id` of the commenter, `photo_id` of the associated photo, and creation timestamp."

A block of code is set as follows:

```
Request:
{
  "userId": "12345",
  "restaurantId": "78566",
  "items": [
    {
      "itemId": "item1",
      "quantity": 4
    },
    {
      "itemId": "item2",
      "quantity": 3
    }
  ],
  "paymentMethod": "credit_card"
}
```

Bold: Indicates a new term, an important word, or words that you see onscreen. For instance, words in menus or dialog boxes appear in **bold**. Here is an example: " The **Photo Upload Service** will expose the following API endpoints."

> **Tips or important notes**
> Appear like this.

Get in touch

Feedback from our readers is always welcome.

General feedback: If you have questions about any aspect of this book, email us at `customercare@packtpub.com` and mention the book title in the subject of your message.

Errata: Although we have taken every care to ensure the accuracy of our content, mistakes do happen. If you have found a mistake in this book, we would be grateful if you would report this to us. Please visit `www.packtpub.com/support/errata` and fill in the form.

Piracy: If you come across any illegal copies of our works in any form on the internet, we would be grateful if you would provide us with the location address or website name. Please contact us at `copyright@packt.com` with a link to the material.

If you are interested in becoming an author: If there is a topic that you have expertise in and you are interested in either writing or contributing to a book, please visit `authors.packtpub.com`.

Share Your Thoughts

Once you've read *System Design Guide for Software Professionals*, we'd love to hear your thoughts! Scan the QR code below to go straight to the Amazon review page for this book and share your feedback.

https://packt.link/r/1-805-12499-4

Your review is important to us and the tech community and will help us make sure we're delivering excellent quality content.

Download a free PDF copy of this book

Thanks for purchasing this book!

Do you like to read on the go but are unable to carry your print books everywhere?

Is your eBook purchase not compatible with the device of your choice?

Don't worry, now with every Packt book you get a DRM-free PDF version of that book at no cost.

Read anywhere, any place, on any device. Search, copy, and paste code from your favorite technical books directly into your application.

The perks don't stop there, you can get exclusive access to discounts, newsletters, and great free content in your inbox daily

Follow these simple steps to get the benefits:

1. Scan the QR code or visit the link below

https://packt.link/free-ebook/9781805124993

2. Submit your proof of purchase

3. That's it! We'll send your free PDF and other benefits to your email directly

Part 1: Foundations of System Design

In this *Part*, you will gain a comprehensive understanding of the fundamental concepts and principles that underpin modern system design. We'll start by exploring the essence of system design, its various types, and its critical importance in today's technology-driven industry.

As we delve deeper, you'll learn about the key principles of distributed systems, which form the backbone of most large-scale applications today. We will cover crucial concepts such as consistency, availability, partition tolerance, latency, durability, reliability, and fault tolerance.

We'll then explore the theoretical foundations that guide system design decisions. You will learn about important theorems and algorithms such as the CAP theorem, PACELC theorem, Paxos and Raft algorithms, and the Byzantine Generals Problem. We'll also introduce you to practical concepts such as consistent hashing, Bloom filters, and HyperLogLog, which are frequently used in large-scale system design.

By the end of this section, you'll have a solid theoretical foundation that will inform your approach to designing and architecting large-scale systems.

This *Part* has the following chapters.

- *Chapter 1, Basics of System Design*
- *Chapter 2, Distributed System Attributes*
- *Chapter 3, Distributed Systems Theorems and Data Structures*

1

Basics of System Design

System design is an essential process in software engineering that involves designing a software system's architecture, components, interfaces, and data management strategies. A well-designed system can improve the system's overall performance, user experience, and security while reducing development costs and time.

This chapter introduces the topic of software system design and talks about the various types and the significance of system design in the software industry. The chapter also discusses system design's impact on software development, maintenance, and overall system performance. By the end of this chapter, you will have a good background in software system design and its importance in software development and have the motivation to dive into this topic even further.

In this chapter, we will cover the following:

- What is system design?
- What are the different types of system design?
- Importance of system design in the industry

What is system design?

Software system design is the process of defining the architecture, components, interfaces, and other characteristics of a system to satisfy specified requirements.

Let's gain an understanding of system design by first delving into the concepts of software systems and distributed software systems.

Software system

A software system is a collection of software components, modules, and programs that work together to perform a specific task or set of tasks. It typically includes a set of interrelated software applications that work together to provide a desired functionality, such as managing data, processing transactions, or delivering a service to end users.

It can be as simple as a single program or as complex as a distributed system that spans multiple computing devices and networks.

A software system is designed, developed, and maintained by software engineers, who use various tools, programming languages, and methodologies to ensure that the system is reliable, scalable, and secure. The software system may also require regular updates and maintenance to keep it functioning properly and to address any issues or bugs that arise over time.

Distributed software system

A distributed software system consists of multiple independent components, processes, or nodes that communicate and coordinate with each other to achieve a common goal. Unlike a centralized software system, where all the components are located on a single machine, a distributed software system is spread across multiple machines, networks, and geographical locations (see *Figure 1.1*).

Each component in a distributed software system is responsible for a specific task or set of tasks, and they work together to achieve a common goal. The components communicate with each other using a variety of communication protocols, such as **remote procedure calls** (**RPCs**), message passing, or publish-subscribe mechanisms.

Distributed software systems are often used in large-scale applications where scalability, availability, and fault tolerance are critical requirements. Examples of distributed software systems include cloud computing platforms, peer-to-peer networks, distributed databases, and **content delivery networks** (**CDNs**).

Designing, developing, and maintaining a distributed software system can be challenging because it requires careful consideration of network communication, data consistency, availability, fault tolerance, and security.

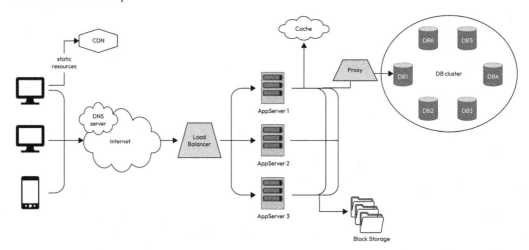

Figure 1.1 – An example of a distributed system

Figure 1.1 shows an example of a distributed system with multiple computer resources and networks. There are user devices (laptops, mobile phones, and tablets) on the left side, which the user interfaces with to submit requests and consume information. These requests then are routed to servers via the DNS server and load balancer to be processed. The server talks to different types of storage systems and databases to fetch the data needed to process and respond to the original request.

Understanding system design

System design is the process of defining the architecture, components, modules, interfaces, and interactions of a software system to meet its functional and non-functional requirements. It involves transforming a set of requirements into a blueprint or a plan that describes how the software system will be structured, implemented, and maintained.

> **Note**
> The goal of software system design is to create a design that is easy to understand, maintain, and extend and that meets the performance, scalability, reliability, and security requirements of the system. The design should also be flexible enough to accommodate changes in the requirements or the environment over time.

The software system design process typically involves the following steps:

1. **Requirements analysis**: Understanding and defining the functional and non-functional requirements of the system. This step also calls for a deeper look into the read-and-write patterns and then designing the system to take advantage of these patterns.

2. **High-level architecture design**: Defining the overall structure of the system (see *Figure 1.2*), including its components, modules, and interfaces.

3. **Detailed design**: Defining the internal structure and behavior of each component and module. This also involves the core algorithms of each component and mechanisms of interactions between components.

4. **User interface design**: Designing the user interface of the system that would interact with the backend services via APIs. This is to be done at a very high level.

5. **API design**: Defining proper APIs, which would enable the user interface or the frontend to interact with the backend services.

6. **Database design**: Designing the data structures and storage mechanisms used by the system. The database could be simple file storage to a relational database such as MySQL or a NoSQL database such as HBase or Cassandra.

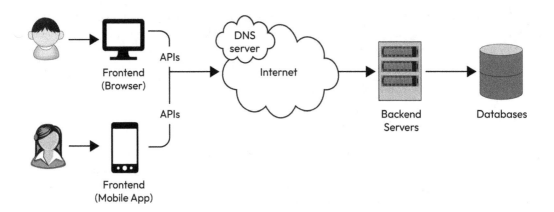

Figure 1.2 – A simple high-level system design diagram of a typical web application

Here, in *Figure 1.2*, we can see an example of a high-level system design architectural diagram, which you can refine as you move on to a more detailed design by adding more components and systems.

The output of the software system design process is a set of design documents, such as architectural diagrams and detailed design documents, enlisting and defining the APIs and user interface prototypes, which serve as a blueprint for implementing the software system.

What are the types of system design?

There are essentially two types of system design: architectural design, also referred to as high-level system design, and detailed design, also referred to as low-level system design.

High-level system design

The key aspects of high-level system design include the following:

- **System architecture**: The overall structure of the system, including its components, relationships, and communication patterns

- **Data flow**: The movement of data through the system, from ingestion to storage and processing

- **Scalability**: The ability of the system to handle increased workloads without significant degradation in performance

- **Fault tolerance**: The capacity of the system to continue functioning despite failures or errors

Let us look at each of these aspects in some more detail.

System architecture

A crucial aspect of high-level system design is defining the overall architecture, which outlines the main components, their relationships, and communication patterns. Some popular architectural patterns include the following:

- **Monolithic**: A single, self-contained application that combines all system components
- **Client-server**: A distributed architecture where clients request services from one or more servers
- **Microservices**: A modular architecture with small, independent services that communicate over a network
- **Event-driven**: A system where components communicate through asynchronous events or messages

When designing a system architecture, consider the following factors:

- **Scalability**: Will the architecture support the system's growth in terms of users, data, and functionality?
- **Maintainability**: How easy will it be to update, debug, or enhance the system?
- **Reliability**: Can the architecture ensure the system's uptime, fault tolerance, and resilience?
- **Latency**: How will the architecture affect the system's response time and performance?

High-level system design focuses on the clarity of system architectural choices. Let us now look into the next aspect, the flow of the data in the system.

Data flow

Understanding how data flows through the system is another essential aspect of high-level system design. A well-designed data flow ensures that the system can ingest, process, store, and retrieve data efficiently. When designing the data flow, consider the following:

- **Data ingestion**: Identify the sources of data and the mechanisms for ingesting it into the system (e.g., APIs, streaming, or batch processing).
- **Data storage**: Determine the appropriate storage solutions for the data, considering factors such as access patterns, query performance, and consistency requirements.
- **Data processing**: Design the processes that transform, analyze, or aggregate the data, considering the necessary compute resources and potential bottlenecks.
- **Data retrieval**: Define how the processed data is accessed by clients or other components, considering latency, caching, and load balancing strategies.

Data flow directly impacts system performance, scale, and usability. Hence, it is important to make the right choices in selecting and/or designing the data flow subsystems.

Scalability

Scalability is a critical aspect of high-level system design, as it determines the system's ability to handle increased workloads without significant degradation in performance. There are two primary types of scalability:

- **Vertical scalability**: Improving performance by adding resources to a single component, such as increasing CPU, memory, or storage
- **Horizontal scalability**: Improving performance by distributing the workload across multiple components or instances, such as adding more servers to a cluster

When designing for scalability, we must consider load balancing, caching, data partitioning, and exploring stateless services. Let us look at the final aspect of high-level system design, that is, fault tolerance.

Fault tolerance

Fault tolerance is the system's ability to continue functioning despite failures or errors in its components. A fault-tolerant system is more reliable and less prone to downtime. Some key strategies for designing fault-tolerant systems include replication, redundancy, graceful degradation, monitoring, and self-healing.

 In short, high-level system design focuses on the high-level architecture of the system and does not delve into the implementation and optimizations. Let us now explore low-level system design.

Low-level system design

Low-level system design focuses on the implementation details of the system's components. This includes selecting appropriate algorithms, data structures, and APIs to optimize performance, memory usage, and code maintainability.

Key aspects of low-level system design include the following:

- **Algorithms**: The step-by-step procedures for performing calculations, data processing, and problem-solving
- **Data structures**: The organization and management of data in memory
- **APIs**: The interfaces that enable communication between different components or services
- **Code optimization**: Techniques to improve code performance, readability, and maintainability

Let us now look into each of these aspects in some more detail.

Algorithms

Algorithms are step-by-step procedures for performing calculations, data processing, and problem-solving in low-level system design. Choosing efficient algorithms is essential to optimize the system's performance, resource usage, and maintainability. When selecting an algorithm, consider the following factors:

- **Time complexity**: The relationship between the input size and the number of operations the algorithm performs
- **Space complexity**: The relationship between the input size and the amount of memory the algorithm consumes
- **Trade-offs**: The balance between time and space complexity, depending on the system's requirements and constraints

Writing an optimized algorithm is a far better choice than leveraging high-end machines in any system. Hence, it is one of the core pillars of a robust system. Algorithms can be optimized by the use of appropriate data structures, which we will cover next.

Data structures

Data structures are used to organize and manage data in memory, impacting the system's performance and resource usage. Choosing appropriate data structures is crucial for low-level system design. When selecting a data structure, consider the following factors:

- **Access patterns**: The frequency and nature of data access, including reads, writes, and updates
- **Query performance**: The time complexity of operations such as search, insertion, and deletion
- **Memory usage**: The amount of memory required to store the data structure and its contents

Some common data structures used in system design include arrays, linked lists, hash tables, trees, and graphs.

APIs

Application programming interfaces (**APIs**) are essential for communication between different components or services in a system. They define the contracts that enable components to interact while maintaining modularity and separation of concerns. When designing APIs, consider the following factors:

- **Consistency**: Ensure that the API design is consistent across all components, making it easy to understand and use
- **Flexibility**: Design the API to support future changes and extensions without breaking existing functionality

- **Security**: Implement authentication, authorization, and input validation to protect the system from unauthorized access and data breaches

- **Performance**: Optimize the API for low latency and efficient resource usage

Having clean and clear APIs is often an enabler to building backward-compatible systems.

Code optimization

Code optimization refers to techniques that improve code performance, readability, and maintainability. In low-level system design, code optimization is essential for ensuring that the system performs well under real-world conditions. Some code optimization techniques include the following:

- **Refactoring**: Restructure the code to improve its readability and maintainability without changing its functionality

- **Loop unrolling**: Replace repetitive loop structures with a series of statements, reducing loop overhead and improving performance

- **Memorization**: Reduce time to recompute results by storing the results of previous calls

- **Parallelism**: Break down tasks into smaller, independent subtasks that can be executed concurrently, reducing overall processing time

We have just provided a flavor of some of the techniques to optimize code. This is a broad topic and has several books and other resources dedicated to it. We would highly recommend you carry out some online research on this topic. Thus, low-level system design focuses on the implementation, interface, and optimization of the system.

We have provided a very high-level overview here of the types of system design. Depending on the stage of the project, you may, as an architect, find yourself dabbling in different forms of system design aspects.

Importance of system design in the industry

Incorporating the right process for system design has several benefits. Some of them are as follows:

- **Clarity of understanding the requirements**: System design enables a clear understanding of requirements, which helps in building the right solution for the problem. This includes identifying the core functionalities, performance, and security requirements of the system.

- **Better collaboration**: System design helps teams to collaborate more effectively, ensuring that everyone involved in the project has a clear understanding of the system's architecture and design. This leads to better communication and coordination among team members and stakeholders.

- **Design reviews and feedback**: Having a system design in place makes it easier for teammates and architects to participate in design reviews to discuss, find issues, and incorporate feedback.

- **High scalability**: Scalability is the ability of a system to handle increasing amounts of data or traffic without compromising performance. System design helps to identify the scalability requirements of the system and design it in a way that can be easily scaled up or down as needed.

- **Performance**: System design ensures that the software solution performs optimally under different loads and usage patterns and avoids performance bottlenecks by thinking ahead of time. It also takes into account factors such as response time, reliability, and availability, which are critical for ensuring user satisfaction.

- **Maintainability**: A well-designed system is easier to maintain and update, reducing the cost of maintenance and improving the system's longevity.

- **Cost-effectiveness**: A well-designed system can be built more efficiently and cost-effectively, as it reduces the risk of errors and rework.

Overall, system design plays a critical role in the development of efficient, effective, and scalable systems that meet the needs of end users and stakeholders.

Practical examples of the importance of system design

The following are practical examples of the importance of software system design in various industries:

- **Finance**: Financial institutions rely heavily on software systems to manage transactions, customer accounts, and other critical operations. Software system design ensures that these systems are secure, reliable, and efficient.

- **E-commerce**: E-commerce platforms require complex software systems to handle large volumes of online transactions and manage inventory, shipping, and customer information. Effective software system design ensures that these platforms are user-friendly, secure, and scalable.

- **Healthcare**: Electronic health records, medical imaging systems, and other healthcare software applications are critical to patient care. Software system design ensures that these applications are reliable, secure, and compliant with regulatory requirements.

- **Manufacturing**: Software systems are used in manufacturing to control production processes, monitor equipment performance, and manage inventory. Effective software system design ensures that these systems are integrated, efficient, and reliable.

- **Transportation**: Software systems are used to manage logistics, track shipments, and optimize delivery routes in the transportation industry. Software system design ensures that these systems are reliable, secure, and able to handle large volumes of data.

In general, software system design is important in any industry where software applications are used to manage complex operations, automate processes, and optimize performance. Effective software system design ensures that these applications are reliable, efficient, and user-friendly, ultimately leading to improved business outcomes.

Summary

In this chapter, we explored the importance of system design and its role in developing software solutions that meet both functional and non-functional requirements. System design involves defining the architecture, components, modules, interfaces, and interactions of a software system. By effectively translating requirements into a well-structured blueprint, system design forms the foundation for successful software development.

We discussed two types of system design: high-level and low-level. High-level system design encompasses critical aspects such as system architecture, data flow, scalability, and fault tolerance. Conversely, low-level system design focuses on implementation details and specific components. Throughout the design process, a set of design documents is produced, serving as valuable blueprints for the actual implementation of the software system.

The significance of system design in the industry cannot be overstated. It promotes a clear understanding of requirements, facilitates collaboration, enables thorough design reviews and feedback, and ensures scalability, performance, and maintainability. Furthermore, well-designed systems contribute to efficient and cost-effective development while reducing the risk of errors and rework. Industries such as finance, e-commerce, healthcare, manufacturing, and transportation particularly benefit from effective software system design, as it empowers them to manage complex operations, automate processes, and optimize performance, ultimately driving improved business outcomes.

Looking ahead to the next chapter, we will delve into the methodologies and techniques employed during the system design process. We will explore practical approaches to translating requirements into well-designed software systems, equipping readers with valuable insights and strategies to enhance their system design capabilities.

2
Distributed System Attributes

Distributed systems have become an integral part of modern computing infrastructure. With the rise of cloud computing and the internet, distributed systems have become increasingly important for providing scalable and reliable services to users around the world. However, designing and operating distributed systems can be challenging due to several factors, including the need for consistency, availability, partition tolerance, and low latency. Other attributes, such as scalability, durability, reliability, and fault tolerance, are critical requirements for any business application catering to a large and diverse demography. A good understanding of these attributes is crucial to designing large and complex systems that address business needs.

In this chapter, we will understand how these distributed system attributes come into play when we think about designing a distributed system. We may need to make appropriate trade-offs among these attributes to satisfy the system requirements.

We will be covering the following concepts in detail in this chapter:

- Consistency
- Availability
- Partition tolerance
- Latency
- Durability
- Reliability
- Fault tolerance
- Scalability

A hotel room booking example

Before we jump into the different attributes of a distributed system, let's set some context in terms of how reads and writes happen.

Let's consider an example of a hotel room booking application (*Figure 2.1*). A high-level design diagram helps us understand how *writes* and *reads* happen:

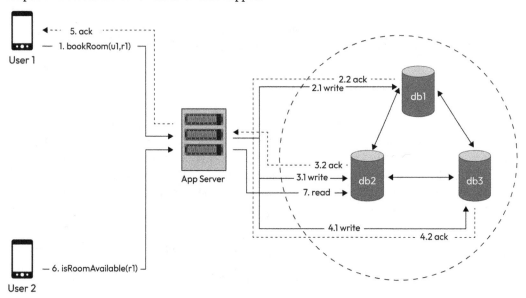

Figure 2.1 – Hotel room booking request flow

As shown in *Figure 2.1*, a user (*u1*) is booking a room (*r1*) in a hotel and another user is trying to see the availability of the same room (*r1*) in that hotel. Let's say we have three replicas of the reservations database (*db1*, *db2*, and *db3*). There can be two ways the writes get replicated to the other replicas: The app server itself writes to all replicas or the database has replication support and the writes get replicated without explicit writes by the app server.

Let's look at the write and the read flows:

Write flow:

User (u1) books a room (r1). The device/client makes an API call to book a room (u1,r1) to the app server. The server writes to one, a few, or all of the replicas.

Read flow:

User (u2) checks the availability of room (r1). The device/client makes an API call in RoomAvailable (r1) to the app server. The server reads from one, a few, or all of the replicas.

Write options:

For write, we have the following options:

- **Serial sync writes**: The server writes to db1 and gets an ack, then writes to db2 and gets an ack, and then writes to db3 and gets an ack. Finally, it acks the client. In this case, the response latency back to the user (u1) would be very high.

- **Serial async writes**: The server writes to db1 and gets an ack. The server asks the client. Asynchronously, the server updates the other two replicas. Write latency is low.

- **Parallel async writes**: The server fires three updates simultaneously, but doesn't wait for all the acks, gets one (or k) acks, and then returns an ack to the client. Latency is low, but thread resource usage is high.

- Write to a messaging service such as Kafka and return an ack to the client. A consumer then picks up the writes and follows any of the aforementioned options. Latency in this case is the lowest. It can support very high writes.

Read options:

For read, we have the following options:

- Read from only one replica

- Read from a quorum number of replicas

- Read from all replicas and then return to the client

Each of these read options comes with consistency trade-offs. For example, if we read from only one replica, the read may be stale in some situations, posing a correctness problem. On the other hand, reading from all replicas and comparing all the values to determine which one is the latest value addresses the correctness problem, but this would be slower. Reading from a quorum number of replicas may be a more balanced approach. We will explore these trade-offs more in the following sections.

We will use this context in understanding the distributed system attributes.

Consistency

Consistency in distributed system design is the idea that all nodes in a distributed system should agree on the same state or view of the data, even though the data may be replicated and distributed across multiple nodes. In other words, consistency ensures that all nodes store the same data and return updates to the data in the same order on being queried for the same updates to the data in the same order.

There are primarily two types of consistency models that can be used in distributed systems:

- Strong consistency

- Eventual consistency

Let's explore the first type of consistency.

Strong consistency

Strong consistency in distributed systems refers to a property that ensures all nodes in the system observe the same order of updates to shared data. It guarantees that when a write operation is performed, any subsequent read operation will always return the most recent value. Strong consistency enforces strict synchronization and order of operations, providing a linearizable view of the system.

To achieve strong consistency, distributed systems employ mechanisms such as distributed transactions, distributed locking, or consensus protocols such as the Raft or Paxos algorithms. These mechanisms coordinate the execution of operations across multiple nodes, ensuring that all nodes agree on the order of updates and maintain a consistent state.

Strong consistency offers a straightforward and intuitive programming model as it guarantees predictable and deterministic behavior. Developers can reason about the system's state and make assumptions based on the order of operations. However, achieving strong consistency often comes at the cost of increased latency and reduced availability as the system may need to wait for synchronization or consensus before executing operations.

A good example of a system that would require a strong consistency model is a banking system. Banking and financial applications deal with sensitive data, such as account balances and transaction histories. Ensuring strong consistency is crucial to avoid discrepancies and to prevent erroneous operations that could lead to financial losses or incorrect accounting.

Eventual consistency

Eventual consistency, on the other hand, is a consistency model that allows for temporary inconsistencies in the system but guarantees that eventually, all replicas or nodes will converge to a consistent state. In other words, it allows updates that are made to the system to propagate asynchronously across different nodes, and eventually, all replicas will agree on the same value.

Unlike strong consistency, where all nodes observe the same order of updates in real-time, eventual consistency relaxes the synchronization requirements and accepts that there may be a period during which different nodes have different views of the system's state. This temporary inconsistency is typically due to factors such as network delays, message propagation, or replica synchronization.

Eventual consistency is often achieved through techniques such as **conflict resolution**, **replication**, and **gossip protocols**. When conflicts occur, such as concurrent updates to the same data on different nodes, the system applies conflict resolution strategies to reconcile the differences and converge toward a consistent state. Replication allows updates to be propagated to multiple replicas asynchronously, while gossip protocols disseminate updates across the system gradually.

The key characteristic of eventual consistency is that given enough time, without further updates or conflicts, all replicas will eventually converge to the same value. The convergence time depends on factors such as network latency, update frequency, and conflict resolution mechanisms.

Eventual consistency offers benefits such as increased availability and scalability and faster response times to the client application. It allows different nodes to continue operating and serving requests, even in the presence of network partitions or temporary failures. It also provides the opportunity to distribute the workload across different replicas, improving system performance.

However, eventual consistency introduces the challenge of dealing with temporary inconsistencies or conflicts. Applications must handle scenarios where different nodes may have different views of the system's state and employ techniques such as conflict resolution, versioning, or reconciliation algorithms to ensure eventual convergence.

The choice of consistency model, whether strong consistency or eventual consistency, depends on the specific requirements of the application. Strong consistency is suitable for scenarios where immediate and strict synchronization is required, while eventual consistency is a trade-off that offers increased availability and scalability at the expense of temporary inconsistencies.

In the hotel room booking example, as shown diagrammatically in *Figure 2.2*, when user (*u1*) books the room, let's say the write goes to only *db1* and then it gets replicated to *db2* and *db3*. While this is being replicated, user (*u2*) makes a call to check if the room (*r1*) is available for booking. The API call may return "true" or "false" depending on whether the write has been replicated to *db2* or not:

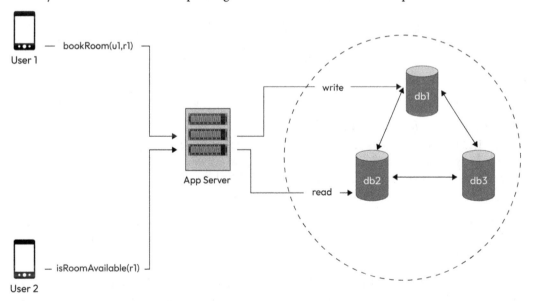

Figure 2.2 – Hotel room booking example to understand consistency

As system designers, we have the option to design strong consistency or eventual consistency. Let's see how we do that.

In this scenario, we have the following:

n = the number of replicas

r = the number of replicas we consider reading from

w = the number of replicas we consider writing to

We talk to all n replicas, but consider w or r number of replicas for evaluation.

Here are our options:

a. w=1, r=3 → strong consistency, fast writes, slow reads

b. w=3, r=1 → strong consistency, slow writes, fast reads

c. w=2, r=2 → strong consistency, writes and reads are both the same pace

d. w=1, r=1 → eventual consistency, fast writes, fast reads

> **Note**
> We always have strong consistency if (r+w) > n; otherwise, it's eventual consistency.

Is eventual consistency okay in this hotel booking use case? The answer may seem trivial at this point. We want to have strong consistency, right? Well, that may not be the case if we consider availability in the mix. Sometimes, we may want eventual consistency as a trade-off to have higher availability. More on this in the next section.

Availability

Availability in distributed system design refers to the ability of a distributed system to provide access to its services or resources to its users, even in the presence of failures. In other words, an available system is always ready to respond to requests and provide its services to users, regardless of any faults or failures that may occur in the system.

In the hotel room booking example, the system can be highly available if the writes and reads happen from only one or a quorum of replicas. This ensures that the user requests will be served by fewer nodes and doesn't require all the nodes to be up. So, in case one or more nodes are in a failed state, the system as a whole is available to take writes and reads.

Achieving high availability in distributed systems can be challenging because distributed systems are composed of multiple components, each of which may be subject to failures such as crashes, network failures, or communication failures.

To ensure availability, distributed systems employ various techniques and strategies, including the following:

- **Redundancy**: Having redundant components or resources enables the system to continue functioning, even if some components fail. Redundancy can be implemented at various levels, such as hardware redundancy (for example, redundant power supplies or network links) and software redundancy (for example, redundant processes or service instances).

- **Replication**: When there is redundancy in the system, we need to replicate the data across these multiple redundant nodes. Replicating data or services across multiple nodes helps ensure that even if one or more nodes fail, others can take over and continue to provide the required functionality. Replication can be done through techniques such as active-passive replication, where one node serves as the primary while others act as backups, or active-active replication, where multiple nodes serve requests simultaneously.

- **Load balancing**: Distributing the workload evenly across multiple nodes helps prevent the overloading of individual nodes and ensures that resources are efficiently utilized. Load balancing mechanisms route incoming requests to available nodes, optimizing resource utilization and avoiding CPU, memory, or I/O bottlenecks that may arise if all requests are served by a small subset of nodes.

- **Fault detection and recovery**: Distributed systems employ mechanisms to detect failures or faults in nodes or components. Techniques such as heartbeating, monitoring, or health checks are used to identify failed nodes, and recovery mechanisms are implemented to restore or replace failed components.

- **Failover and failback**: Failover mechanisms automatically redirect requests from a failed node or component to a backup or alternative node. Failback mechanisms restore the failed node or component once it becomes available again.

By implementing these techniques, distributed systems can provide high availability, reducing the impact of failures or disruptions and ensuring continuous access to services or resources. However, achieving high availability often involves trade-offs, such as increased complexity, resource overhead, or potential inconsistencies or performance compromises, which need to be carefully considered based on the specific requirements of the system.

Understanding partition tolerance

Before we take a look at partition tolerance, let's understand what a partition (or network partition) is.

Network partition

A network partition in distributed systems refers to a situation where a network failure or issue causes a subset of nodes or components to become disconnected or isolated from the rest of the system, forming separate groups or partitions. In other words, the network partition divides the distributed system into multiple disjoint segments that cannot communicate with each other.

Network partitions can occur due to various reasons, such as network failures, hardware malfunctions, software bugs, or unintentional consequences due to planned actions such as network configuration changes or network attacks. An example is shown in *Figure 2.3*, where the *db2* node is isolated and can't communicate with the other two nodes. When a network partition happens, the nodes on one side of the partition can no longer send messages or exchange information with the nodes on the other side. In this example, any writes to *db1* or *db3* can't propagate updates to *db2*. In this scenario, if a user's read requests land on *db2*, it may serve stale data:

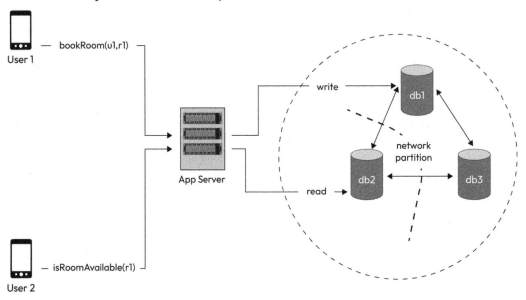

Figure 2.3 – Network partition scenario

Hence, the existence of network partitions poses challenges for distributed systems because it disrupts the communication and coordination between nodes. Nodes within the same partition can continue to interact and operate normally, but they are unable to reach nodes in other partitions. This can lead to inconsistencies, conflicts, and challenges in maintaining system properties such as consistency, availability, and fault tolerance.

Network partitions, as shown in *Figure 2.3*, can have different characteristics and implications based on their duration and severity. They can be transient, lasting for a short period, and resolving automatically once the network issue is resolved. Alternatively, partitions can be long-lasting or permanent if network connectivity cannot be restored.

Partition tolerance

Now, partition tolerance (or network partition tolerance) is a property of distributed systems that refers to the system's ability to continue functioning despite network failures or network partitions.

In a distributed system that is designed with network partition tolerance, the system can continue to operate despite these network failures. Nodes that are isolated due to a network partition can still function independently and serve their clients, while the rest of the system continues to operate as usual. When we discuss the CAP theorem (in the next chapter), we will see how the system behaves when they are partition-tolerant and what trade-off we need to make between consistency and availability.

Partition tolerance holds immense significance in distributed systems, particularly in scenarios where high availability is crucial, such as cloud computing, distributed databases, and large-scale distributed applications. By embracing partition tolerance, distributed systems can ensure uninterrupted operations and graceful degradation in the face of network failures or partitions, thereby enhancing robustness and fault tolerance.

Latency

Latency is the time delay between the initiation of a request and the response to that request in a distributed system design. In other words, it is the time it takes for data to travel from one point to another in a distributed system.

Latency is an important metric in distributed system design because it can affect the performance of the system and the user experience. A system with low latency will be able to respond to requests quickly, providing a better user experience, while a system with high latency may experience delays and be perceived as slow or unresponsive.

Latency can be influenced by a variety of factors, including the distance between nodes in the system, network congestion, processing time at each node, and the size and complexity of the data being transmitted.

Reducing latency in distributed systems can be challenging as it involves optimizing various aspects of the system. Some techniques and strategies to mitigate latency include the following:

- **Network optimization**: Optimizing network infrastructure, such as using high-speed connections, reducing network hops, and minimizing network congestion, can help reduce latency.
- **Caching**: Implementing caching mechanisms at various levels, such as in-memory caching or **content delivery networks** (**CDNs**), can improve response times by serving frequently accessed data or content closer to the user.
- **Data localization**: Locating data or services closer to the users or consumers can help reduce latency. This can be achieved through data replication, edge computing, or utilizing content distribution strategies.

- **Asynchronous communication**: Using asynchronous communication patterns, such as message queues or event-driven architectures, can decouple components and reduce the impact of latency by allowing parallel processing or non-blocking interactions.
- **Performance tuning**: Optimizing system configurations, database queries, algorithms, and code execution can help improve overall system performance and reduce latency.

It's important to note that while minimizing latency is desirable, it may not always be possible to eliminate it entirely. Distributed systems often operate in environments with inherent network delays, and achieving extremely low latency may come at the cost of other system properties, such as consistency or fault tolerance. Therefore, the appropriate trade-offs should be made based on the specific requirements and constraints of the distributed system.

Durability

Durability in a distributed system design is the ability of the system to ensure that data stored in the system is not lost due to failures or errors. It is an important property of distributed systems because the system may be composed of multiple nodes, which may fail or experience errors, potentially leading to data loss or corruption.

To achieve durability, a distributed system may use techniques such as **replication** and **backup**. Data can be replicated across multiple nodes in the system so that if one node fails, the data can still be retrieved from another node. Additionally, backup systems may be used to store copies of data in case of a catastrophic failure or disaster.

Durability is particularly important in systems that store critical data, such as financial or medical records, as well as in systems that provide continuous service, such as social media or messaging platforms. By ensuring durability, we can ensure that the system is reliable and that data is always available to users.

It is important to note that durability is closely related to other properties of distributed systems, such as consistency and availability. Achieving high durability may require trade-offs with other properties of the system. Hence, we must carefully balance these factors when designing and implementing a distributed system.

Reliability

Reliability in distributed systems means that the system can consistently provide its intended functionality, despite the occurrence of various failures and errors such as hardware failures, network issues, software bugs, and human errors. A reliable distributed system ensures that data and services are always available, accessible, and delivered promptly, even in the face of these challenges.

Reliability is a crucial aspect of distributed systems, which are composed of multiple interconnected nodes or components working together to achieve a common goal. Achieving reliability in distributed systems requires the implementation of various techniques, such as redundancy, fault tolerance, replication, load balancing, and error handling. These techniques help prevent, detect, and recover from failures, ensuring that the system remains operational and consistent in its behavior.

Fault tolerance

Fault tolerance in distributed systems means that the system continues functioning correctly in the presence of component failures or network problems. It involves designing and implementing a system that can detect and recover from faults automatically, without any human intervention.

To achieve fault tolerance, distributed systems employ various techniques, such as redundancy, replication, and error detection and recovery mechanisms. Redundancy involves duplicating system components or data to ensure that if one fails, another can take its place without disrupting the overall system. Replication involves creating multiple copies of data or services in different locations so that if one location fails, others can still provide the required service.

Error detection and recovery mechanisms involve constantly monitoring the system for errors or failures and taking appropriate actions to restore its normal functioning. For example, if a node fails to respond, the system may try to communicate with another node or switch to a backup component to ensure uninterrupted service.

Overall, fault tolerance ensures that distributed systems can continue to provide their services even in the presence of failures or errors, increasing their reliability and availability.

Scalability

Scalability in distributed systems refers to the ability of a system to handle an increasing workload as the number of users or size of data grows, without sacrificing performance or reliability. It involves designing and implementing a system that can efficiently and effectively handle larger amounts of work, either by adding more resources or by optimizing existing resources.

In distributed systems, there are primarily two types of scaling – vertical scaling and horizontal scaling (as shown in *Figure 2.4*):

- Vertical scaling (scaling up)
- Horizontal scaling (scaling out):

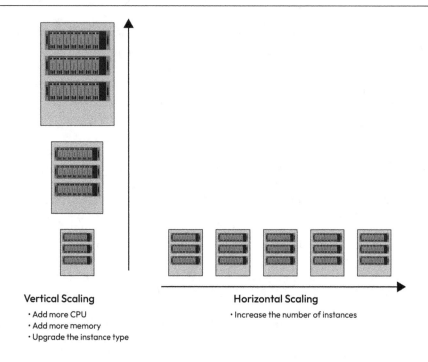

Vertical Scaling
- Add more CPU
- Add more memory
- Upgrade the instance type

Horizontal Scaling
- Increase the number of instances

Figure 2.4 – Vertical scaling versus horizontal scaling

Let's look into these two types of scaling in detail.

Vertical scaling

Vertical scaling, also known as scaling up, involves increasing the capacity of an individual node/instance or resource within the node. It typically involves upgrading hardware components, such as increasing the CPU power, adding more memory, or expanding storage capacity. Vertical scaling focuses on improving the performance of a single node to handle increased workloads or demands.

Here are some of the advantages of vertical scaling:

- **Simplicity**: It generally requires minimal changes to the existing system architecture or software

- **Cost-effectiveness for smaller workloads**: Vertical scaling can be a more cost-effective approach for systems with relatively lower workloads as it eliminates the need for managing and maintaining a large number of nodes

However, there are limitations to vertical scaling:

- **Hardware limitations**: There is a limit to how much a single node can be upgraded in terms of CPU power, memory, or storage. Eventually, a point is reached where further upgrades become impractical or too expensive.

- **Single point of failure**: Since the system relies on a single node, in case that node fails, the entire system may become unavailable.

In the next section we will discuss horizontal scaling.

Horizontal scaling

Horizontal scaling, also known as scaling out, involves adding more nodes or instances to the distributed system to handle increased workloads or demands. It focuses on distributing the workload across multiple nodes, allowing for parallel processing and improved system capacity.

Here are the advantages of horizontal scaling:

- **Increased capacity and performance**: By adding more nodes, the system can handle a higher volume of requests or data processing, leading to improved performance and responsiveness.

- **Fault tolerance**: With multiple nodes, the system becomes more resilient to failures. If one node fails, the others can continue to operate, ensuring the availability of the system.

Horizontal scaling also has considerations and challenges:

- **Distributed coordination**: Distributing the workload across multiple nodes requires effective coordination and synchronization mechanisms to ensure consistent and correct results.

- **Data consistency**: Maintaining data consistency can be more challenging in distributed systems compared to vertical scaling, where data resides on a single node. Techniques such as distributed transactions or eventual consistency need to be employed to handle data across multiple nodes.

It's worth noting that vertical and horizontal scaling are not mutually exclusive, and they can be combined to achieve the desired scalability and performance goals for a distributed system. Often, a combination of both approaches is used to scale different components or layers of the system effectively. The choice between vertical and horizontal scaling depends on factors such as the specific requirements of the system, workload patterns, cost considerations, and the ability to effectively distribute and coordinate the workload across multiple nodes.

To ensure scalability, distributed systems must be designed with a modular and loosely coupled architecture, which allows for easy addition or removal of components as needed. Additionally, the system must be able to dynamically adjust its resource allocation in response to changing workload demands.

Overall, scalability is crucial for distributed systems as it allows them to handle increasing workloads and continue providing their services effectively and efficiently.

Summary

In this chapter, we explored the critical aspects of distributed system design, including consistency, availability, partition tolerance, latency, durability, reliability, fault tolerance, and scalability. We examined the importance of achieving consistency to ensure all nodes in a distributed system observe the same updates in the same order. Additionally, we discussed different consistency models, such as strong consistency and eventual consistency, and their implications for system design.

Ensuring high availability in distributed systems poses challenges due to the potential for system failures. We examined techniques such as redundancy, replication, and failover, all of which can be employed to improve system availability. We also emphasized the significance of considering partition tolerance, which refers to the system's ability to continue functioning despite network failures or partitions. Designing distributed systems with effective partition tolerance capabilities is crucial for maintaining system reliability and uninterrupted operation.

Latency, another critical factor in distributed system design, was explored in detail. We discussed how latency, the time delay between request initiation and response, can be influenced by various factors, including network congestion, node distances, data size, and processing time. Understanding and managing latency is essential for optimizing system performance and ensuring timely and efficient communication between distributed system components.

Durability and reliability are crucial considerations in distributed systems. We examined techniques such as replication, backup, redundancy, and fault tolerance, all of which contribute to achieving data durability and consistent system functionality, even in the presence of failures or errors. Finally, we explored scalability in distributed systems, discussing the benefits and considerations of vertical scaling (scaling up) and horizontal scaling (scaling out) to handle increasing workloads without sacrificing performance or reliability.

By gaining a comprehensive understanding of these fundamental aspects of distributed system design, you will be better equipped to make informed decisions and address challenges in your own distributed system architectures.

In the next chapter, we will dive deeper into understanding distributed systems by discussing the relevant theorems and data structures.

3

Distributed Systems Theorems and Data Structures

Various theorems, algorithms, and data structures play a crucial role in the design and implementation of distributed systems. By exploring these concepts, we aim to provide a solid foundation for understanding and tackling the intricacies of building reliable, scalable, and fault-tolerant distributed systems.

We will dive deeper into a collection of essential theorems that form the theoretical underpinnings of distributed systems. These theorems provide formal proofs and insights into various aspects of distributed computing, such as consensus protocols, distributed algorithms, and fault tolerance. We'll examine classical theorems such as the CAP theorem, the PACELC theorem, the FLP impossibility result, and the **Byzantine generals problem** (**BGP**), among others. These theorems serve as guiding principles for reasoning about the limitations and possibilities of distributed systems. With a solid grasp of the foundational theorems, we'll shift our focus to various techniques and data structures that are used in distributed systems.

We will cover the following topics in this chapter:

- CAP theorem
- PACELC theorem
- The Paxos and Raft algorithms
- BGP
- FLP impossibility theorem
- Consistent hashing
- Bloom filters
- Count-min sketch
- HyperLogLog

Let's begin by exploring the CAP theorem.

CAP theorem

The **CAP theorem**, also known as **Brewer's theorem**, is a fundamental principle in distributed systems. It states that a distributed system cannot simultaneously provide consistency, availability, and partition tolerance all at once. The acronym CAP represents the three properties.

In distributed systems, network partitions are an inevitable occurrence due to various reasons, such as hardware failures, network outages, or even routine maintenance. These partitions lead to nodes being split into isolated groups, disrupting the normal flow of communication. Consequently, the system faces a crucial decision in the face of such partitions: prioritizing between consistency and availability. On one hand, if a system opts for **availability and partition tolerance (AP)**, it continues to function despite the partition but may sacrifice consistency, meaning all nodes might not have the same data at the same time. On the other hand, prioritizing **consistency and partition tolerance (CP)** ensures all nodes have the same data, but this might come at the cost of the system's availability, potentially leading to downtimes or reduced functionality during partitions. It's important to note that compromising between these three aspects – consistency, availability, and partition tolerance – is fundamentally impossible due to the CAP theorem.

In a distributed system, especially those spread across multiple locations or relying on the internet, network partitions are inevitable. These partitions can be caused by various factors, such as hardware failures, network outages, routing issues, or even large-scale disasters. Given this inevitability, a system must be designed to handle such partitions. Ignoring partition tolerance means assuming a perfect, fault-free network, which is unrealistic in practical scenarios.

Essentially, when a distributed system encounters a network partition, where nodes are separated into isolated groups, it must choose between consistency and availability. In other words, during a partition, a system can prioritize availability and partition tolerance, sacrificing consistency (AP), or it can prioritize consistency and partition tolerance, sacrificing availability (CP).

It's crucial to understand that the CAP theorem doesn't imply an all-or-nothing sacrifice of properties in every situation. Instead, it highlights the inherent trade-offs that distributed systems face, requiring designers to make conscious decisions based on their specific system requirements and priorities.

Different distributed systems may opt for different trade-offs based on factors such as the nature of the application, user needs, and expected network conditions. For instance, in scenarios where strict consistency is vital, such as financial transactions, a CP system may be preferred. Conversely, in scenarios prioritizing high availability and responsiveness, such as web applications, an AP system may be chosen:

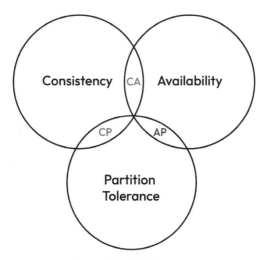

Figure 3.1 – CAP theorem

The CAP theorem, as shown in *Figure 3.1*, establishes a foundation for comprehending the challenges and design considerations in distributed systems. It assists system architects in making informed decisions regarding consistency, availability, and partition tolerance based on the unique requirements of their systems.

PACELC theorem

The **PACELC theorem**, an extension of the CAP theorem, considers the trade-offs in distributed systems when confronted with network partitions. Let's take a look at what PACELC stands for:

- **Partition tolerance (P)**: Partition tolerance refers to a distributed system's ability to function and provide services even when network partitions or communication failures occur. It means the system can tolerate the loss of network connectivity between different nodes.

- **Availability (A)**: Availability ensures that every request made to the distributed system eventually receives a response, regardless of the state of individual nodes or network partitions. The focus is on providing timely responses, even if it means sacrificing strong consistency.

- **Consistency (C)**: Consistency entails all nodes in a distributed system agreeing on the current state of the system. Strong consistency guarantees that every read operation sees the most recent write operation. However, achieving strong consistency in a distributed system often leads to increased latency or reduced availability.

- **Else (E)**: The "E" in PACELC represents the "else" scenario, where there is no network partition. In this case, a trade-off arises between latency and consistency.

- **Latency (L):** Latency refers to the time delay between initiating a request and receiving a response. In some cases, optimizing for low latency may require relaxing consistency guarantees or reducing availability.

- **Consistency Level (C):** The consistency level indicates the desired or provided level of consistency in a distributed system. It can vary based on application requirements or the design choices made by system architects.

The PACELC theorem offers a more nuanced perspective compared to the CAP theorem. It states that in the presence of a network partition, we must choose between consistency and availability (as in the CAP theorem). However, even without a network partition, a trade-off emerges between latency and consistency. Different distributed systems may prioritize high consistency at the expense of increased latency or opt for lower latency with eventual consistency. *Figure 3.2* captures how PACELC is different from the CAP theorem:

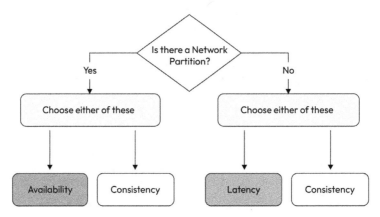

Figure 3.2 – PACELC theorem

Understanding the PACELC theorem helps system architects and designers make informed decisions based on the specific requirements and priorities of their distributed systems while considering both network partitions and the impact on latency and consistency.

In the realm of distributed systems, ensuring consensus among multiple nodes is a fundamental challenge. In this section, we will discuss two seminal algorithms that help with distributed consensus: **Paxos** and **Raft**.

Paxos

Paxos is a consensus algorithm that was introduced by Leslie Lamport in 1990, and with its elegant design, Paxos has become a cornerstone of modern distributed systems, serving as the foundation for numerous applications, including databases and distributed storage systems.

At its core, Paxos aims to enable a distributed system to agree on a single value, even in the presence of failures and network delays. The "single value" in the context of the Paxos algorithm refers to a specific piece of data or a decision that the nodes in a distributed system need to agree upon. This consensus is crucial for ensuring that the system behaves consistently and reliably, particularly in scenarios where multiple nodes might propose different values or updates. It achieves this by employing a protocol that allows a group of nodes to reach a consensus on a proposed value through a series of communication rounds.

Key components

To understand Paxos, we need to familiarize ourselves with its key components:

- **Proposers**: These are the nodes that are responsible for initiating the consensus process. A proposer suggests a value to be agreed upon and broadcasts this proposal to the other nodes in the system.

- **Acceptors**: Acceptors are the nodes that receive proposals from proposers. They play a crucial role in the protocol by accepting proposals and communicating their acceptance to other nodes.

- **Learners**: Learners are the final recipients of the agreed-upon value. Once consensus is reached, learners acquire the value and take appropriate actions based on it.

Now, let's dive into the protocol itself.

Protocol steps

The Paxos protocol proceeds through a series of steps, as shown in *Figure 3.3*, which can be summarized as follows:

- **Prepare phase**: A proposer selects a unique proposal number and sends a prepare request to a majority of acceptors. Acceptors respond with the highest-numbered proposal they have accepted.

- **Accept phase**: If the proposer receives responses from a majority of acceptors, it proceeds to the accept phase. The proposer sends an accept request, along with its proposal number and value, to the acceptors.

- **Consensus reached**: If the majority of the acceptors accept the proposal, consensus is reached, and the value is chosen. The learners are then informed of the chosen value:

Figure 3.3. shows how the Paxos algorithm operates in practice.

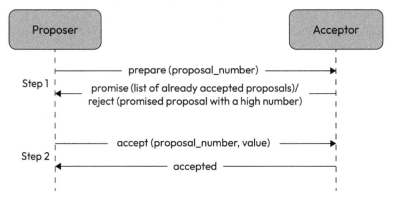

Figure 3.3 – The Paxos algorithm

While Paxos provides a robust mechanism for achieving consensus in distributed systems, it is not without its challenges:

- **Fault tolerance**: Paxos accounts for failures and network delays by tolerating the absence of some nodes and ensuring progress despite potential disruptions.

- **Scalability**: The performance of Paxos can be impacted by the number of nodes involved. As the system scales, coordination and communication overhead may increase.

- **Complexity**: Paxos is renowned for its elegance, but it can be challenging to understand and implement correctly. Careful attention must be given to ensure all participants adhere to the protocol's requirements.

Paxos variants and optimization techniques

Over the years, researchers and practitioners have introduced various variants and optimization techniques to improve the efficiency and understandability of Paxos. Some notable advancements are shown here:

- **Multi-Paxos**: Multi-Paxos extends the basic Paxos protocol to allow for continuous consensus on multiple values without repeating the prepare and accept phases. It reduces the overhead of repeated agreement rounds and enables faster agreement on subsequent values.

- **Fast Paxos**: Fast Paxos introduces an optimization to reduce the number of messages required for consensus. It allows a proposer to bypass the traditional prepare phase by proposing a value directly to the acceptors, thereby reducing the latency in reaching a consensus.

- **Simple Paxos**: Simple Paxos aims to simplify the original Paxos protocol by combining the prepare and accept phases into a single round. This reduces the number of message exchanges that are required and enhances the protocol's understandability.

Real-world use cases

Paxos finds applications in various distributed systems that require fault-tolerant consensus. Here are some notable use cases:

- **Distributed databases**: Paxos is commonly used in distributed databases to ensure consistency and durability across replicas. It allows database nodes to agree on the order of committed transactions and handle failures gracefully.

- **Distributed filesystems**: Filesystems such as Google's GFS and Hadoop's HDFS leverage Paxos to maintain the consistency and availability of file metadata across multiple nodes. Paxos ensures that all replicas agree on the state of the filesystem, even in the presence of failures.

- **Replicated state machines**: Paxos serves as the foundation for implementing replicated state machines, where a cluster of nodes agrees on a sequence of commands or operations to maintain consistency across replicas. This enables fault tolerance and replication in systems such as distributed key-value stores and consensus-based algorithms.

Hence, we can say that Paxos provides a robust and widely adopted solution for achieving consensus in distributed systems. Its elegance and versatility have made it a key building block for numerous applications. By understanding this conceptual overview, as well as key components, protocol steps, challenges, and real-world use cases of Paxos, software engineers can effectively leverage this consensus algorithm to design and build reliable distributed systems.

In the next section, we will explore Raft and its applications.

Raft

Raft, a consensus algorithm introduced by Diego Ongaro and John Ousterhout in 2013, provides a simplified and intuitive approach to distributed consensus. With its emphasis on understandability and ease of implementation, Raft has gained popularity as an alternative to more complex consensus algorithms such as Paxos.

Raft aims to enable a distributed system to agree on a single, consistent state, even in the presence of failures. It achieves this by dividing the consensus problem into three subproblems: leader election, log replication, and safety. By tackling these subproblems, Raft simplifies the coordination and communication required among nodes to reach consensus. Raft differs from Paxos in that it has a designated "leader."

Key components

To understand Raft, let's explore its key components:

- **Leader**: Raft operates under the assumption that the system has a designated leader. The leader is responsible for managing the consensus process and coordinating the replication of log entries across other nodes.

- **Followers**: Followers are passive nodes that replicate the leader's log and respond to incoming requests. They rely on the leader for guidance in the consensus process.

- **Candidate**: When a leader fails or a new leader needs to be elected, nodes transition to the candidate state. Candidates initiate leader election by requesting votes from other nodes in the system.

As you can see, most of the key components of Raft are similar to Paxos but Raft has a designated leader. Now, let's dive into the protocol's steps.

Protocol steps

The Raft protocol proceeds through a series of steps, as shown in Figure 3.4, which can be summarized as follows:

- **Leader election**: When a system starts or detects the absence of a leader, a new leader needs to be elected. Nodes transition to the candidate state and send out `RequestVote` messages to other nodes. A candidate becomes the leader if they receive votes from a majority of nodes.

- **Log replication**: The leader is responsible for receiving client requests, appending them to its log, and replicating the log entries to followers. Followers apply received log entries to their state machines to maintain consistency across the system.

- **Safety and consistency**: Raft ensures safety and consistency by enforcing specific rules, such as the "Append Entries" rule and the "Voting" rule. These rules prevent inconsistencies and guarantee that only the most up-to-date log entries are committed. See *Figure 3.4*.

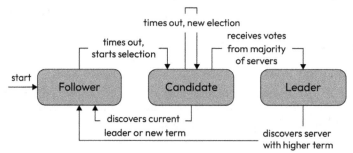

Figure 3.4 – Raft protocol steps

Challenges and considerations

While Raft simplifies the consensus problem, there are some considerations to keep in mind:

- **Leader availability**: The availability of a leader is crucial for the progress of the system. If the leader becomes unavailable, a new leader must be elected promptly to avoid interruptions in the consensus process.

- **Scalability**: As the system scales and the number of nodes increases, the communication and coordination overhead can become a performance bottleneck. Proper optimization techniques and configuration adjustments are necessary to ensure scalability.

- **Fault tolerance**: Raft provides fault tolerance by allowing nodes to detect leader failures and initiate leader elections. The algorithm ensures that the system can continue to make progress, even in the presence of failures.

Practical applications

The following are some practical applications of the Raft algorithm with examples of how it is used in all major companies companies:

- **Distributed databases – Amazon DynamoDB**: Distributed databases require a consensus algorithm such as Raft to maintain consistency and durability across replicas. Raft ensures that all nodes agree on the order of committed transactions and handle failures gracefully. Amazon DynamoDB, a highly scalable and managed NoSQL database service, employs distributed consensus mechanisms such as Raft to ensure data consistency and availability across its distributed infrastructure.

- **Distributed filesystems – Google File System (GFS)**: Distributed filesystems rely on consensus algorithms to maintain consistency and availability of file metadata across multiple nodes. Raft is employed to ensure that all replicas agree on the state of the filesystem, even in the presence of failures. GFS, which is used by Google for storing large-scale distributed data, utilizes consensus algorithms such as Raft to achieve fault tolerance, data consistency, and replication across its distributed filesystem.

- **Cluster coordination and service discovery – Apache ZooKeeper**: Distributed systems often require coordination and consensus among nodes for tasks such as leader election and service discovery. Raft is used in cluster coordination frameworks such as Apache ZooKeeper to provide a reliable and consistent coordination service. ZooKeeper leverages Raft to ensure consensus on critical system metadata, such as leader election, configuration changes, and distributed locks, enabling fault-tolerant coordination among distributed services.

- **Consensus-based algorithms – Google Spanner**: Consensus-based algorithms such as Paxos and Raft are fundamental in building distributed systems that require strong consistency guarantees. Google Spanner, a globally distributed database, utilizes Raft to ensure consistency and fault tolerance across its globally distributed replicas. Raft helps maintain consensus on the order of operations and transaction commits, ensuring data integrity and consistency in Spanner's distributed architecture.

- **Cloud infrastructure management –Netflix's Chaos Automation Platform (ChAP)**: Cloud infrastructure management platforms, such as Netflix's ChAP, require coordination and agreement among distributed components for tasks such as resource allocation, fault tolerance, and auto-scaling. Raft can be employed to provide consensus and coordination among the different components of the platform, ensuring that resource management decisions are made consistently across the distributed infrastructure.

These are just a few examples of how the Raft algorithm is practically applied within top tech companies. Raft's simplicity, understandability, and fault-tolerant properties make it a preferred choice for distributed consensus in various scenarios, from distributed databases and filesystems to cluster coordination and infrastructure management.

The Raft algorithm provides a reliable consensus protocol when dealing with crash failures, but it does not address more subtle failure scenarios, such as nodes that behave erroneously or maliciously. This leads us to a more complex consensus problem known as BGP.

BGP

BGP is a classic thought experiment in the area of reliability and fault tolerance in distributed systems. The problem illustrates the challenges of achieving reliable consensus when some components are unreliable or behaving unexpectedly.

Imagine a group of generals of the Byzantine Empire of Rome, in around 300 CE, camped with their troops around an enemy city. The generals can use only a messenger to communicate with each other. To win the battle, all the generals must agree upon a common plan of action. Some of the generals could be traitors who can confuse the loyal generals.

The loyal generals need a way to reliably agree upon a coordinated plan of action, even in the presence of these traitorous generals spreading false information. This is a non-trivial problem because of the following reasons:

- The generals can only communicate through a messenger, which can fail or be intercepted.

- Some fraction of the generals may be traitors who will deliberately try to confuse the loyal generals and prevent consensus.

- The loyal generals do not know which generals are loyal and which are traitors. All generals seem identical from the outside.

- The traitorous generals may collude together to prevent consensus among the loyal generals.

There are a few basic requirements for a solution to BGP:

- If the majority of the generals are loyal, a solution must allow the loyal generals to eventually reach a consensus

- The loyal generals must be able to tolerate up to f traitors, where $f < (n-1)/2$ and n is the total number of generals

- The loyal generals must be able to make decisions in a reasonable amount of time

The key challenge is reliably achieving consensus in a distributed system where some arbitrary number of components may be faulty or adversarial. This problem illustrates the complexities involved and serves as an instructive starting point for the development of fault-tolerant distributed systems and algorithms.

There are a few common solutions to BGP:

- **Voting algorithms**: The generals vote on the proposed plan of action. If a threshold of votes is reached (for example, a 2/3 majority), then that plan is chosen. Traitorous generals can cast faulty votes, but so long as less than 1/3 of generals are traitors, the loyal generals can still reach a consensus.

- **Multi-round signing**: The generals propose a plan and sign it to vote for it. If a plan gets signatures from 2/3 of the generals, it is chosen. Any general who does not sign a chosen plan is identified as a traitor. This is done over multiple rounds to weed out traitors.

- **Quorum systems**: The generals are divided into quorums, where each quorum must have a majority of loyal generals. Each quorum votes on a plan independently. If a plan gets a majority vote in every quorum, it is chosen. The intersection of all quorums ensures a plan has the majority support of loyal generals.

- **Timeouts and questioning**: Generals propose a plan and vote for it. If a general does not vote within a timeout, it is marked as a suspect. Generals can also question other generals about their votes and remove generals who give conflicting answers.

- **Randomization**: Generals propose a plan with a random nonce (number used once). Traitorous generals cannot know the correct nonce in advance. If a general's vote has the correct nonce, it is likely loyal. Randomization makes it harder for traitors to interfere successfully.

In general, solutions involve some form of voting, identifying faulty or traitorous generals, and redundancy to tolerate a minority of failures while ensuring consensus among the majority of loyal generals. The challenge lies in designing algorithms that are efficient, resilient, and able to make forward progress even under adversarial conditions.

The Byzantine fault

The term **Byzantine fault** refers to the unpredictable and unreliable behavior exhibited by some nodes in the system, comparable to the traitorous generals in BGP. Byzantine faults can include a variety of issues, such as software bugs, hardware failures, or malicious attacks, which can result in a node failing in arbitrary ways.

These faults are particularly challenging to manage as they can produce false or contradictory information, making it difficult for the system to identify and isolate the problematic nodes.

Byzantine fault tolerance

To counter BGP, a system needs to implement **Byzantine fault tolerance** (**BFT**). This is a property of a system that allows it to function correctly and reach consensus, even when some nodes fail or act maliciously. In other words, a system is BFT if it can still provide its services accurately to users, despite the Byzantine faults.

One of the earliest solutions for BFT was proposed by Lamport, Shostak, and Pease. They proposed a voting system where each node sends its value (or "vote") to every other node. Each node then decides on the majority value. However, the protocol only works if less than a third of the nodes are faulty. This solution requires every node to communicate with every other node, resulting in high computational and communication overheads.

Modern BFT

Modern systems often implement variations of BFT algorithms, such as the **Practical Byzantine Fault Tolerance** (**PBFT**) protocol or variants of it. PBFT reduces the communication overhead and allows for more efficient consensus mechanisms in large networks. These solutions have been instrumental in the operation of modern distributed systems, including blockchain technologies such as Bitcoin and Ethereum.

BGP is a critical challenge in system design, highlighting the difficulty of achieving consensus in a distributed system, especially when some nodes may behave unpredictably or maliciously. The study of BFT and the creation of algorithms to achieve it is crucial to the design and implementation of reliable, resilient distributed systems and networks.

As system designers, understanding BGP, its implications, and the methods to achieve BFT equips us to build robust and reliable distributed systems. The world is increasingly reliant on such systems – from digital currencies and distributed databases to large-scale computing clusters – making BGP more relevant than ever.

FLP impossibility theorem

The **FLP impossibility theorem**, named after **Fischer, Lynch,** and **Paterson**, who proved it in 1985, states that it is impossible to design a totally asynchronous distributed system that can reliably solve the consensus problem in the presence of even one failure.

The consensus problem requires that all processes in a distributed system eventually agree on a single value, given some initial set of proposed values. For example, in BGP, the generals need to reach a consensus on a plan of attack.

The key assumptions in the FLP impossibility proof are as follows:

- **Asynchrony**: There are no bounds on message delays or process speeds. Processes communicate by sending messages but have no shared clock.

- **Process failures**: Up to f out of n processes may fail by crashing, where $0 < f < n$.

- **Finite steps**: Processes take a finite number of steps and messages have a finite size.

Under these conditions, the FLP theorem proves that there is no deterministic algorithm that can ensure all correct processes reach consensus. This impossibility holds even if only one process fails. The FLP impossibility states that it is impossible to simultaneously achieve all three of the desirable properties in an asynchronous distributed system, such as fault tolerance, agreement, and termination. We can design systems that can only achieve two of these properties, as depicted in *Figure 3.5*:

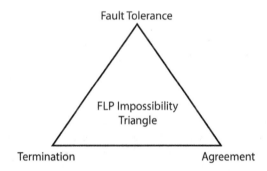

Figure 3.5 – FLP impossibility triangle

The intuition behind the proof is that in an asynchronous system, processes have no way to differentiate between a slow process and a crashed process. A live process may appear dead to other processes due to long message delays, creating ambiguity that prevents consensus from being reached reliably.

The implications of the FLP impossibility theorem are significant. It shows that purely asynchronous distributed systems are fundamentally limited in what problems they can solve reliably. Approaches to overcome the impossibility result typically involve one of the following:

- Introducing synchronous assumptions, such as known bounds on message delays
- Using probabilistic algorithms that can reach a consensus with a high probability
- Adding randomness or a clock synchronization mechanism
- Implementing a leader election or coordinator role

Overall, the FLP impossibility theorem helps explain why distributed consensus remains a complex and challenging problem and highlights the trade-offs involved in designing distributed algorithms and systems. It serves as a theoretical bound on what is computationally possible in distributed systems under certain assumptions.

Hence, the FLP theorem proves that totally asynchronous distributed systems cannot deterministically solve the consensus problem in the presence of even one process failure. This fundamental impossibility has significant ramifications and suggests strategies for overcoming the limitations.

Now that we've talked about various theorems, let's shift gears to understand some of the techniques and data structures that are widely used in designing distributed systems. A good understanding of these concepts will help us build a foundation for future chapters.

Consistent hashing

Consistent hashing is a technique that's used in distributed systems to efficiently distribute data across multiple nodes while minimizing the need for data reorganization or rebalancing when nodes are added or removed from the system. It provides a scalable and fault-tolerant approach to handling data distribution.

In traditional hashing techniques, such as modulo hashing, the number of nodes or buckets where data can be stored is fixed. When nodes are added or removed, the hash function that's used to map data keys to nodes changes, requiring a significant amount of data to be remapped and redistributed across the nodes. This process can be time-consuming, resource-intensive, and disruptive to the system's availability.

The key problem that consistent hashing addresses is the scalability and fault tolerance of distributed systems. It aims to minimize the impact of adding or removing nodes from the system by ensuring that only a fraction of the data needs to be remapped when the number of nodes changes. This property makes consistent hashing particularly useful in large-scale systems where nodes are frequently added or removed, such as **content delivery networks** (**CDNs**) or distributed databases.

In consistent hashing, a hash function is used to map data keys and node identifiers to a common hash space. The hash space is typically represented as a ring, forming a circular continuum. Each node in the distributed system is also assigned a position on the ring based on its identifier. To store or retrieve data, the same hash function is applied to the data key, mapping it to a position on the ring. Moving clockwise on the ring from the key's position, the first node encountered is responsible for storing or handling that particular data. This node is known as the "owner" or "responsible node" for that data.

Let's consider an example to understand this better. There are user API requests and we need to assign these requests to servers. Let's assume we have four servers (`server0`, `server1`, `server2`, and `server3`) in this example. *Figure 3.6* shows the hash ring where we will be mapping the requests and the servers. We need to choose a hash function that takes the server IDs and generates points on this hash ring. We will use the same hash function and pass the request IDs to generate points to be mapped on this hash ring. Let's call the hash function `h()`.

Here, servers get mapped to points in the ring, like this:

`h(server0) = s0`

`h(server1) = s1`

`h(server2) = s2`

`h(server3) = s3`

The requests get mapped to points in the ring, like this:

`h(req_id1) = r1`

`h(req_id2) = r2`.... and so on.

Now, to determine which server takes on which request, we take the point (let's say `r1`) corresponding to the request (`req_id1`) and walk clockwise to find the first server point that's encountered, which in this case is `s1` (corresponding to `server1`):

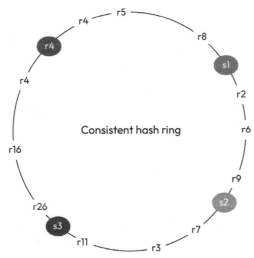

Figure 3.6 – Simple consistent hashing ring

This initial design still has a couple of issues:

The distribution of the requests to the servers may not be even. In *Figure 3.6*, it appears that all four points corresponding to the four servers are placed equidistant. But that may not always be true.

When one server goes down, the next server (walking in a clockwise direction) will have to bear the load of the failed server as well. That may be too much for the server. As a result, this next server may also go down due to excessive load, which can create a chain reaction that causes all the servers to go down.

Let's enhance the consistent hashing technique to fix these issues. *Figure 3.7* shows an enhanced version of the consistent hashing ring:

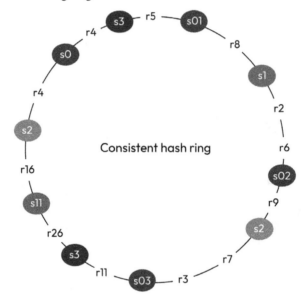

Figure 3.7 – Enhanced consistent hashing ring

In this enhanced approach, we use extra dummy points for the servers and place them in different positions in the hash ring. In this example, let's consider creating three extra dummy points for each server. To do this, we must use three extra hash functions (h1 (), h2 (), and h3 ()) and use them to create three extra points in the hash ring corresponding to each server.

So, for server0, we have the following:

```
h(server0)  = s0
h1(server0) = s01
```

```
h2(server0) = s02
h3(server0) = s03
```

For `server1`, we have the following:

```
h(server1)  = s1
h1(server1) = s11
h2(server1) = s12
h3(server1) = s13
```

Similarly, we generate points for `server2` and `server3`.

As you can see, when using this enhanced consistent hashing technique, we have a better distribution of nodes (the original point and the dummy points for the servers) so that the requests are assigned to the server more evenly. Also, when a server goes down, the four corresponding points will go down, but since the distribution of points corresponding to other servers is well distributed, the requests will be almost evenly distributed among the surviving servers, not just one. This helps prevent the cascading failure issue we mentioned earlier with the simple consistent hashing ring.

So, in conclusion, by using a hash ring and mapping data keys and node identifiers to a common hash space, consistent hashing provides a way to consistently assign data to nodes in a scalable and fault-tolerant manner. When a node is added or removed, only a portion of the data needs to be remapped to the neighboring nodes, limiting the amount of data movement required and minimizing disruption to the system.

Bloom filters

A **Bloom filter** is a space-efficient probabilistic data structure that is used to test whether an element is a member of a set. It provides a fast and memory-efficient way to perform membership queries. The primary advantage of a Bloom filter is its ability to quickly determine if an element is "possibly" in a set, with a small probability of false positives.

The Bloom filter is based on the concept of hash functions and bit arrays. It consists of a bit array of a fixed size and a set of hash functions. Initially, all the bits in the array are set to 0. To add an element to the filter, it is hashed by the set of hash functions, and the corresponding bits in the array are set to 1. To check the membership of an element, it is hashed again using the same hash functions, and the corresponding bits are examined. If any of the bits are 0, the element is not in the set. However, if all the bits are 1, it is only probable that the element is in the set, as there is a chance of false positives.

Let's understand the Bloom filter by looking at an example. Let's say we're checking the availability of usernames for a website. We need to give a unique username to a new user.

Here, we have the following aspects to consider:

- The number of hash functions is $k = 2$
- The bitcount (the size of the bit array) is $m = 10$

The two hash functions are h1() and h2(), both of which generate a number between 0 and 9.

Initialize the bit array via a = [0, 0, 0, 0, 0, 0, 0, 0, 0, 0].

Now, there are 3 usernames, john, dave, peter.

h1("john") = 5; h2("john") = 8,

h1("dave") = 3; h2("dave") = 5,

h1("peter") = 2; h2("peter") = 9

Mark bits for "john" in array [0, 0, 0, 0, 0, 1, 0, 0, 1, 0]

Mark bits for "dave" in array [0, 0, 0, 1, 0, 1, 0, 0, 1, 0]

Mark bits for "peter" in array [0, 0, 1, 1, 0, 1, 0, 0, 1, 1]

Now, let's check whether a username is available or not.

Check "donald" in bloom filter:

h1("donald") = 3; h2("donald") = 7

On checking these two bit positions, we see that donald is for sure not in the existing set (and hence its available)

Check "sarah" in bloom filter h1("sarah") = 3; h2("sarah") = 8

Checking these two bit positions, we see that sarah may not be available.

We can see from the above example that sarah was actually available, but bloom filter said that it's not available (a false positive), but donald was available and it did say that it's indeed available (no false negative).

Bloom filters have several use cases. Let's take a look at a few:

- **Membership testing**: Bloom filters can efficiently test if an element is a member of a set, such as a URL in a web crawler or a word in a spell checker, without the need to store the entire set
- **Caching**: Bloom filters can be used in caching systems to determine if a requested item is present in the cache, thereby avoiding expensive cache lookups

- **Database optimization**: Bloom filters can be utilized to filter out unnecessary disk reads by quickly identifying data that is likely not present in a database, reducing I/O operations

- **Network routing**: Bloom filters can assist in routing decisions by indicating potential destinations for a given packet or message

- **Duplicate elimination**: Bloom filters can help identify duplicates in large datasets, reducing storage requirements and improving efficiency

It is important to note that while Bloom filters provide efficient and fast membership queries, they have a probability of providing false positives. If a Bloom filter reports that an element is present in a set, there is a chance it might not be. However, false negatives (indicating that an element is not present when it is) are not possible.

To mitigate false positives, the size of the bit array and the number of hash functions used can be adjusted. Increasing the size of the bit array reduces the probability of false positives, but it increases memory requirements. Similarly, increasing the number of hash functions decreases the probability of false positives, but it increases computational overhead.

Hence, a Bloom filter is a space-efficient data structure that provides a probabilistic way to test membership in a set. It finds applications in scenarios where quick and memory-efficient membership queries are required, with an acceptance for a small probability of false positives.

Count-min sketch

Count-min sketch is a probabilistic data structure that's used to estimate the frequency of elements in a data stream. It provides an approximate representation of the frequency distribution of elements while using a small amount of memory. Count-min sketch is particularly useful in scenarios where memory is limited or when processing large-scale data streams in real time.

Count-min sketch consists of a two-dimensional array of counters, with the number of rows and columns determined by the desired accuracy and error rate. When an element is encountered in the data stream, multiple hash functions are applied to determine the positions in the array to increment the corresponding counters. By using multiple hash functions, collisions are distributed, and the frequency of elements is estimated across different counters.

Let's consider an example to understand this better. We have a stream of four alphabets (A, B, C, and D) coming to our system and we need to be able to determine the count of these alphabets at any given time.

For this, we will use five hash functions (`depth, d=5`) and an array length of (`w`) `=10`.

The values that correspond to each of the letters when they're passed through the five hash functions are shown in *Figure 3.8(a)*:

	A	B	C	D
h1()	0	1	0	8
h2()	2	3	4	3
h3()	4	4	3	5
h4()	8	7	7	7
h5()	9	2	9	1

Figure 3.8(a)

Initially, all the values for the two-dimensional array state are zeros. This is before any of the letters have been streamed. The initial state is shown in *Figure 3.8(b)*:

	width (w)										
	0	1	2	3	4	5	6	7	8	9	
	0	0	0	0	0	0	0	0	0	0	h1()
	0	0	0	0	0	0	0	0	0	0	h2()
Depth (d)	0	0	0	0	0	0	0	0	0	0	h3()
	0	0	0	0	0	0	0	0	0	0	h4()
	0	0	0	0	0	0	0	0	0	0	h5()

Figure 3.8(b)

Now, let's assume the stream of letters comes as (A, A, B, D, B, A, A, D, B, B, ...).

Here are the steps we must follow:

When the first letter is received, which is A the corresponding hash function values are {0, 2, 4, 8, 9}, as depicted in *Figure 3.8(a)*. What we need to do is increment the cell corresponding to the **h1() row** and **0th** column by 1 . Similarly, we must increment the cell corresponding to (**h2() row, 2nd column**), (**h3() row, 4th column**), (**h4() row, 8th column**) and (**h5() row, 9th column**) by 1. The resultant two-dimensional array looks as follows:

	width (w)										
	0	1	2	3	4	5	6	7	8	9	
	1	0	0	0	0	0	0	0	0	0	h1()
	0	0	1	0	0	0	0	0	0	0	h2()
Depth (d)	0	0	0	0	1	0	0	0	0	0	h3()
	0	0	0	0	0	0	0	0	1	0	h4()
	0	0	0	0	0	0	0	0	0	1	h5()

Figure 3.8(c)

For the second alphabet, **A**, we do what we did previously – that is, increment the cells corresponding to (**h1() row** and **0th column**), (**h2() row, 2nd column**), (**h3() row, 4th column**), (**h4() row, 8th column**) and (**h5() row, 9th column**) by 1. The resultant two-dimensional array looks as follows:

							width (w)				
		0	1	2	3	4	5	6	7	8	9	
Depth (d)	2	0	0	0	0	0	0	0	0	0	h1()
	0	0	2	0	0	0	0	0	0	0	h2()
	0	0	0	0	2	0	0	0	0	0	h3()
	0	0	0	0	0	0	0	0	2	0	h4()
	0	0	0	0	0	0	0	0	0	2	h5()

Figure 3.8(d)

For the third alphabet, B, we do the same thing again but consider the hash values of B, which are {1, 3, 4, 7, 2}. So, we must increment the cells corresponding to (**h1() row** and **1st column**), (**h2() row, 3rd column**), (**h3() row, 4th column**), (**h4() row, 7th column**) and (**h5() row, 2nd column**) by 1. The resultant two-dimensional array looks as follows. Notice that the cell corresponding to (**h3() row, 4th column**) has a collision – both A and B increment the cell value:

							width (w)				
		0	1	2	3	4	5	6	7	8	9	
Depth (d)	2	1	0	0	0	0	0	0	0	0	h1()
	0	0	2	1	0	0	0	0	0	0	h2()
	0	0	0	0	3	0	0	0	0	0	h3()
	0	0	0	0	0	0	0	1	2	0	h4()
	0	0	0	0	0	0	0	0	0	2	h5()

Figure 3.8(e)

Similarly, for the fourth alphabet, D, we must consider the hash values of D, which are {8, 3, 5, 7, 1}. Here, we must increment the cells corresponding to (**h1() row** and **8th column**), (**h2() row, 3rd column**), (**h3() row, 5th column**), (**h4() row, 7th column**) and (**h5() row, 1st column**) by 1. The resultant two-dimensional array looks like this:

							width (w)				
		0	1	2	3	4	5	6	7	8	9	
Depth (d)	2	1	0	0	0	0	0	0	0	0	h1()
	0	0	2	2	0	0	0	0	0	0	h2()
	0	0	0	0	3	1	0	0	0	0	h3()
	0	0	0	0	0	0	0	2	2	0	h4()
	0	0	0	0	0	0	0	0	0	2	h5()

Figure 3.8(f)

Now, if we want to count how many times A appeared in the stream, we must figure out the minimum of the cell values that correspond to the row and column for A. In *Figure 3.8(f)*, the cell values that correspond to (**h1() row** and **0th column**), (**h2() row, 2nd column**), (**h3() row, 4th column**), (**h4() row, 8th column**), and (**h5() row, 9th column**) are (2, 2, 3, 2, 2). The minimum of these 5 values is 2. So, the count of A is 2.

Similarly, we can count how many times **B** appeared by finding the minimum of the cell values that correspond to (**h1() row** and **1st column**), (**h2() row, 3rd column**), (**h3() row, 4th column**), (**h4() row, 7th column**), and (**h5() row, 2nd column**), which is (1, 2, 3, 2, 1) = 1. Hence the count of B is 1.

The accuracy of count-min sketch depends on the number of counters and the number of hash functions used. Increasing the number of counters improves accuracy but also increases memory usage. Similarly, using more hash functions reduces collision probabilities and improves accuracy, but it also introduces additional computational overhead.

Count-min sketch finds applications in various areas, including the following:

- **Frequency estimation**: Count-min sketch can estimate the frequency of elements in a data stream, such as counting the number of times a word appears in a text corpus or tracking the popularity of items in online shopping

- **Traffic analysis**: It can be used in network traffic analysis to estimate the number of packets or flows associated with specific protocols or network addresses

- **Web analytics**: Count-min sketch can approximate the frequency of website visits, clicks, or user interactions, allowing efficient analysis of web traffic

- **Distributed systems**: Count-min sketch is valuable in distributed systems for collecting statistics and monitoring key metrics, such as tracking the frequency of requests across different nodes

- **Data stream processing**: It enables approximate counting and frequency estimation in real-time data streams, where the entire dataset cannot be stored in memory

It's important to note that count-min sketch provides an approximate representation of frequencies and is susceptible to overestimation due to collisions. However, it offers a trade-off between memory usage and accuracy, making it a valuable tool in scenarios where precise frequency counts are not required, and memory constraints are a concern.

HyperLogLog

HyperLogLog is a probabilistic algorithm that's used for estimating the cardinality (or the number of distinct elements) of a set with very low memory usage. It was introduced by Philippe Flajolet and is particularly useful when dealing with large datasets or when memory efficiency is a concern. The HyperLogLog algorithm approximates the cardinality of a set by using a fixed amount of memory, regardless of the size of the set. It achieves this by exploiting the properties of hash functions and probabilistic counting.

The basic idea behind HyperLogLog is to hash each element of the set and determine the longest run of zeros in the binary representation of the hash values. The length of the longest run of zeros is used as an estimation of the cardinality. By averaging these estimations over multiple hash functions, a more accurate cardinality estimate can be obtained.

Let's understand this by considering an example. The problem statement is, *"We need to find the count of unique visitors for a website."*

What are some naive solutions here? We can maintain a hashmap with `key` as the unique user ID and `value` to count how many times a particular user visited the website. This works fine for a low-scale website, but as the website scales up in terms of the number of visitors, the memory footprint grows linearly. So, for one billion visitors, we need 1 GB if we represent each visitor just by 1 byte. Let's explore how HyperLogLog comes to the rescue here.

We need to use randomness here. Let's use a hash function to convert a username into a binary number and assume it's perfectly random. Also, the same username will yield the same hash value. We'll assume we have a perfect hash function that provides a complete random hash and converts the value into a binary representation.

Let's say we have 1 billion users. We need at least 30 bits to represent them:

```
u1(1,000,000,000) -> 111011100110101100101000000000
```

Now, let's say we have the following:

```
hash("John_1275") = 111011100110101100101000001100

hash("David.raymond23") = 111011100110101100001000000010

hash("Sarah1978") = 100011100110101100101000000001

hash("John") -> 111011100110101100101000001100
```

Let's reframe the problem: *"We need to count the unique number of random binary numbers."*

Let's understand how we can do this by using an analogy of flipping a coin – we need to flip the coin until we get a T. Think about getting a sequence – `H, H, H, T`. The probability of getting this is very hard, right?

That's a ½ * ½ * ½ * ½ = `1/16` probability on average. So, that also means that to get `H H H T`, we must flip the coin `16` times.

To extend this, if someone shows that the largest streak of leading heads (`H`) they got was L, this means that, approximately, they flipped the coin `2^(L+1)` times. In the preceding example, L=3, so `2^(3+1) = 16`.

Let's get back to our binary numbers (hash of usernames):

```
hash("John_1275") = 1110111001101011001010000011100
```

```
hash("David.raymond23") = 1110111001101011000010000000010
```

```
hash("Sarah1978") = 1000111001101011001010000000001
```

```
hash("John") = 1110111001101011001010000011100
```

Instead of leading Hs, we will use ending 0s. In the preceding sample of four usernames, the longest subsequence of 0s at the end is 2, so we likely have `2^(2+1) = 8 visitors`.

In a small sample set, this isn't correct, but at a large scale, does it become accurate? Even at a high scale, we can see the accuracy problem because there could be outliers. If there is one bad hash, it will screw up the estimate.

To increase the accuracy, we can follow these steps:

1. Split the incoming hash(usernames) randomly into k buckets.
2. Calculate the number of max ending 0s in each of these buckets.
3. Store `"ending 0 counters"` in these k buckets.
4. Calculate L= average of these k counters. Instead of taking the arithmetic mean, we can take the harmonic mean (this is why it's called HyperLogLog). The harmonic mean is better at ignoring outliers than the arithmetic mean.
5. The arithmetic mean of N numbers is `(n1, n2, ...) = (n1+n2+n3+....)/N`.
6. The harmonic mean of N numbers is `(n1, n2, ...) = N/(1/n1+1/n2+1/n3...)`.
7. Estimate the final number of unique visitors via = `2^(L+1)`.

Calculating the exact cardinality of a multiset requires an amount of memory proportional to the cardinality, which is impractical for very large datasets. HyperLogLog uses significantly less memory than this, at the cost of obtaining only an approximation of the cardinality.

HyperLogLog provides a relatively small memory footprint compared to traditional methods for exact counting, such as storing each element in a set.

We can use the following code to estimate the space of the HyperLogLog counter:

```
2^(L+1) = 1 billion visitors
```

```
L = log2(1000000000) = at max 30 ending 0's
```

```
log2(30) = 5 bits
```

5 bits can represent the number of ending 0s, so let's say a byte. So, even if we use `k =10` counters, it will be 10 bytes.

HyperLogLog has found applications in various domains where cardinality estimation is important, such as database systems, network traffic analysis, web analytics, and big data processing. It is widely used in distributed systems and data streaming scenarios where memory is limited, and fast approximate cardinality estimation is required.

Summary

In this chapter, we embarked on a deep exploration of the essential theorems that serve as the foundation for distributed systems. These theorems offer formal proofs and valuable insights into different facets of distributed computing, including consensus protocols, distributed algorithms, and fault tolerance. By studying these theorems, we gained a comprehensive understanding of the limitations and possibilities that are inherent in distributed systems. Additionally, we talked about some probabilistic data structures that are commonly employed in distributed systems, further expanding our knowledge in this domain.

The first part of this chapter focused on examining classical theorems that are integral to understanding distributed systems. We explored prominent theorems such as the CAP theorem, which delves into the trade-offs between consistency, availability, and partition tolerance. The PACELC theorem, another key theorem, provides insights into the behavior of systems when network partition doesn't happen, but there is a trade-off between consistency and latency.

Next, we covered some important topics, such as the Paxos and Raft algorithms, which are consensus algorithms that are essential for fault-tolerant distributed systems. These algorithms provide a means to achieve agreement among multiple nodes despite potential failures or network partitions.

After that, we dove into the intricacies of BGP, which addresses the challenges that are posed by malicious actors in distributed systems. Understanding this problem is crucial for designing resilient and secure distributed systems. Additionally, we explored the FLP impossibility theorem, which establishes the fundamental limitations of achieving consensus in an asynchronous system with even a single faulty process. This theorem highlights the inherent challenges of ensuring fault tolerance in distributed systems.

Finally, we delved into various techniques and data structures that are employed in distributed systems. We discussed consistent hashing, a technique for distributing data across multiple nodes in a scalable and load-balanced manner. Bloom filters, another data structure, allow for efficient probabilistic set membership testing. Count-min sketch, on the other hand, offers an approximate frequency counting mechanism, which is useful for tracking events in large-scale distributed systems. Lastly, we explored HyperLogLog, a probabilistic algorithm that allows us to estimate the cardinality in sets with minimal memory usage.

With a good understanding of these theorems, algorithms, and data structures, we can start designing and implementing various system components in the next chapters.

Part 2: Core Components of Distributed Systems

In this *Part*, we'll explore the essential building blocks that form the backbone of modern distributed systems. We'll dive deep into the core components that enable scalability, reliability, and performance in large-scale applications.

This section will equip you with a deep understanding of the core components used in building robust, scalable distributed systems.

This section will equip you with a deep understanding of the core components used in building robust, scalable distributed systems. By the end of this *Part*, you'll be well-prepared to make informed decisions about which components to use in your system designs.

- *Chapter 4, Distributed Systems Building Blocks: DNS, Load Balancers, and Application Gateways*
- *Chapter 5, Design and Implementation of System Components – Databases and Storage*
- *Chapter 6, Distributed Cache*
- *Chapter 7, Pub/Sub and Distributed Queues*

4

Distributed Systems Building Blocks: DNS, Load Balancers, and Application Gateways

In this chapter, we shift our focus to the essential building blocks that are instrumental in crafting robust, scalable, and efficient systems. Mastering the intricacies of the **Domain Name System (DNS)**, **load balancers**, and **application gateways** allows for a granular, bottom-up approach to system design, a complement to the theoretical principles discussed in previous chapters.

The knowledge gained here will not only deepen your understanding of system architecture but will also provide you with practical skills that are useful in real-world applications. From ensuring global connectivity with DNS to optimizing server performance with load balancers, and further securing your applications through application gateways, this chapter equips you to tackle complex design challenges head-on.

We will cover the following topics in this chapter:

- Exploring DNS
- Load balancers
- Application gateways

Let us first understand what a DNS is.

Exploring DNS

The DNS maps human-friendly domain names to machine-readable IP addresses. It provides the translation service between domain names and IP addresses. When a user enters a domain name in the browser, the browser needs to find the corresponding IP address to complete the request. It does this by querying the DNS infrastructure. The DNS works transparently in the background. Users are unaware of the domain name to IP mapping performed by DNS. When the browser obtains the IP address from DNS, it forwards the user's request to the destination web server at that IP address.

In short, DNS performs the crucial function of translating domain names that users type into their browsers to IP addresses that computers use to locate websites and resources. This translation is performed seamlessly, allowing users to access websites using easy-to-remember names. This is shown in *Figure 4.1*.

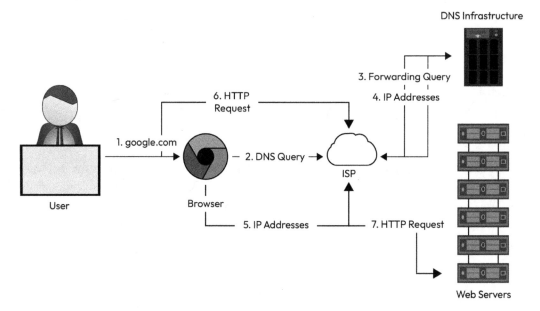

Figure 4.1: Basic architecture of DNS

The DNS is comprised of architectural concepts such as name servers, resource records, caching, and hierarchical arrangement of DNS servers. Let us now look at each of these concepts:

- **Name servers**: The DNS is not a single server, but a network of numerous servers. DNS servers that respond to user queries are called name servers.

- **Resource records (RRs)**: The DNS database stores mappings in units called RRs. There are different RR types that store different kinds of information.

Some common RR types are as follows:

- **A records**: These map hostnames to IP addresses

- **NS records**: These map domain names to authoritative name servers

- **CNAME records**: These map alias hostnames to canonical hostnames

- **MX records**: These map domains to mail servers

- **Caching**: The DNS uses caching at multiple levels to reduce latency. Caching reduces the load on the DNS infrastructure since it handles queries for the entire internet.

- **Hierarchy**: DNS name servers are arranged hierarchically, allowing DNS to scale to its enormous size and query load. The hierarchical structure manages the entire DNS database.

In short, DNS performs its function through a network of distributed name servers, a database of resource records, caching at multiple levels, and a hierarchical structure. These details, as shown in *Figure 4.2*, allow DNS to provide fast, scalable translation of domain names to IP addresses for the entire internet.

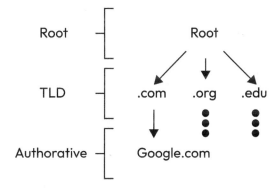

Figure 4.2: DNS name server hierarchy

As seen in the figure, there are different types of DNS name servers. Name servers at different hierarchies are a critical aspect of the DNS infrastructure. Let us understand the different types of server hierarchy:

- **Root-level name servers**: These servers receive queries from local resolvers. They maintain name servers for top-level domains such as `.com`, `.edu`, and `.us`. For example, for `google.com`, root servers will return servers for the `.com` top-level domain.

- **Top-level domain (TLD) name servers**: These servers hold the IP addresses of authoritative name servers for their domain. For the `.com` TLD, they would return the authoritative servers for `google.com`.

- **Authoritative name servers**: These are the organization's actual DNS servers that provide the IP addresses of their web and application servers.

To summarize, we can say that the DNS resolver starts the query process. Root servers point to TLD servers, which point to authoritative servers that finally resolve domain names to IP addresses. This hierarchy allows DNS to scale for the entire internet.

Here are the steps:

1. Resolvers initiate user queries.

2. Root servers point to TLD servers.

3. TLD servers point to authoritative servers.

4. Authoritative servers provide final IP addresses.

5. The hierarchy enables DNS to scale for the internet.

Let us now understand how DNS querying works.

DNS querying

DNS queries basically come in two flavors: **iterative** and **recursive**. These are the main ways a computer finds out what IP address corresponds to a website name such as www.google.com. Let's dig into what each type is all about.

Iterative queries

In an Iterative query, your computer, through its local DNS resolver, does all the legwork. It goes from asking the root name servers at the top of the DNS chain, down to the TLD servers (e.g., .com, .org), and then directly asks the authoritative name servers for the website you want to visit.

Your computer directly talks to the following:

- Root servers

- TLD servers

- Website's authoritative name servers

Your computer walks through these steps one by one.

Figure 4.3 shows how an iterative query works.

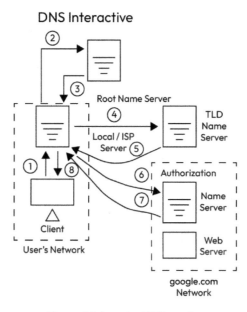

Figure 4.3: Iterative DNS queries

Next, let us understand how the DNS recursive queries work.

Recursive queries

Recursive queries are a bit different. Your computer asks its local DNS resolver for the website's IP address. The local resolver then takes over and does all the asking, moving from the root servers to the TLD servers, and finally to the authoritative servers for the website.

You ask your local resolver for the IP. The local resolver then talks to the following:

- Root servers
- TLD servers
- Website's authoritative name servers

The resolver then gives you back the IP address you need.

Figure 4.4 shows how recursive queries work.

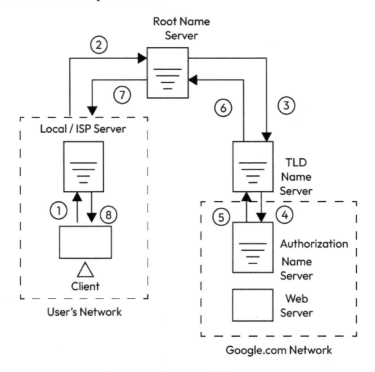

Figure 4.4: Recursive DNS queries

In summary, with iterative queries, the local resolver goes through the DNS hierarchy itself. With recursive queries, the local resolver recursively asks higher-level servers that forward the query down the hierarchy until the final IP address is obtained.

> **Can you guess which mechanism is preferred to reduce the load on the DNS infrastructure?**
>
> Hint: it is not recursive. Recursive DNS queries will require an increase in memory to maintain the recursion structure, whereas iterative does not need that. This implies that recursion will get slower over time as more state needs to be maintained.

When you type a website name, you almost instantaneously get the results of the web page. This slick interface is made possible due to **DNS caching**, which we will discuss in the next section.

Caching

Caching means temporarily storing frequently accessed RRs. An RR is a unit of data that binds a name to a value in the DNS database.

Caching provides two main benefits:

- It reduces response time for users by providing answers locally instead of querying the DNS hierarchy
- It decreases network traffic by avoiding unnecessary queries up the DNS hierarchy

Caching at multiple levels can significantly reduce the load on the DNS infrastructure since it handles queries for the entire internet.

There are many places where caching can be implemented:

- In the user's browser
- In the operating system
- In the local name server within the user's network
- In the **internet service provider (ISP)** resolvers

Therefore, caching DNS resource records at different levels allows answers to be provided locally. This reduces response times for users and decreases traffic in the DNS system, especially when caching is implemented at multiple levels.

DNS is a distributed system and like any other distributed system, it implements some form of scalability, reliability, and consistency, which we will learn in the next section.

Scalability, reliability, and consistency in DNS

When we talk about any large-scale system, especially something as vast as DNS, there are three buzzwords that often come up: *scalability*, *reliability*, and *consistency*. These are like the three pillars that hold up any robust system. Let's dive into how DNS checks off these crucial boxes.

Scalability

The DNS achieves high scalability through its hierarchical design with root, top-level domain, and authoritative name servers distributed worldwide. Roughly 1,000 replicated instances of 13 root-level servers handle the initial user queries. They point to TLD servers, which, in turn, point to authoritative servers managed by individual organizations. This hierarchical structure divides the load between different levels, allowing DNS to scale to serve billions of daily queries for the entire internet.

The distributed nature of DNS also contributes to its scalability. Redundant replicated DNS servers are located strategically around the globe to serve user requests with low latency. If a DNS server becomes overloaded or unavailable, other servers can respond to queries, further enhancing the scalability of the system.

Reliability

The DNS demonstrates high reliability through various mechanisms. Caching at multiple levels – in browsers, operating systems, local networks, and ISP resolvers – provides answers to users even if some DNS servers are temporarily unavailable.

The redundant replicated DNS servers distributed worldwide also improve reliability by serving requests with low latency.

While DNS uses the **User Datagram Protocol** (**UDP**) for most queries and responses, which is inherently unreliable, it compensates through techniques such as resending requests if no response is received. This allows DNS to achieve acceptable levels of reliability for its core function.

Consistency

Though DNS achieves high performance by sacrificing strong consistency, it does provide eventual consistency through techniques such as time-to-live for cached records and lazy propagation of updates across the DNS infrastructure. Updates within the DNS system can take from seconds to days to be reflected across all servers.

However, performance remains the higher priority in DNS design. Even though there are techniques to ensure consistency, the acceptable level of consistency is one that does not degrade performance significantly.

This covers the basics of DNS and we will now explore the second basic building block (i.e., load balancers).

Load balancers

Load balancing distributes workload across multiple computing resources, such as servers, CPUs, hard drives, and network links, to achieve optimal resource utilization, maximize throughput, minimize response time, and avoid overload. *Figure 4.5* shows a load balancer connected to clients on one end and a pool of server machines on the other end.

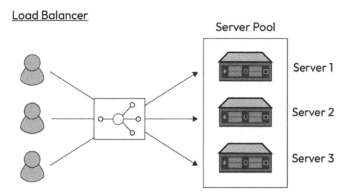

Figure 4.5: Load balancer

The following are some key points about load balancing:

- It allows the spread of a huge amount of traffic across multiple servers so that no single server gets overloaded.

- It improves fault tolerance by failing over (i.e., retrying the failed requests from failed or slowed-down servers to functioning servers).

- It increases overall service availability since requests can still be serviced by the functioning servers even if some servers fail.

- It can enable a graceful degradation of performance during periods of high load instead of a complete failure of the system. As load increases, response time may increase but the system remains operational.

- It enables horizontal scaling by adding more servers into the pool to handle higher loads.

Load balancing is achieved through dedicated hardware devices called load balancers or software solutions that distribute incoming traffic among multiple servers. A load balancer sits in front of the servers and monitors the load on each server, forwarding incoming requests to the server that is least busy.

Placing load balancers

Load balancers can be placed at various points in a system's architecture, as seen in *Figure 4.6*, to distribute load and achieve scalability. This includes placing them between clients and frontend servers, between different tiers of a multi-tier system, and potentially between any two services with multiple instances.

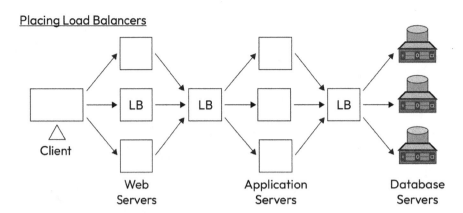

Figure 4.6: Placing load balancers

Placing load balancers between clients and servers allows incoming client requests to be distributed to multiple server instances, which is common for scaling web servers. Load balancers can also be used between layers of a multi-tier system, such as between web servers and application servers, and between application servers and database servers. This enables each layer to be scaled independently, based on its needs.

Distributing traffic at different points in the system also isolates layers from each other, protecting upper layers if a lower layer becomes overloaded or has an outage. The load balancers can detect faults and failures at one layer and redirect traffic around them, keeping the overall system operational.

In summary, the flexibility of load balancers allows them to be strategically placed at critical points in a system's architecture to distribute load, improve scalability, and provide high availability. Load balancers sit at the intersection of where traffic enters and exits different system tiers, serving as a control point for optimizing resource utilization across the system.

Advantages of load balancers

Load balancers are not just leveraged for distributing load and requests, but can improve the availability and health of the systems:

- Load balancers can implement health checks to determine whether backend servers are available or not. This prevents clients from sending requests to unavailable servers.

- Load balancers can terminate TLS (Transport Layer Security)connections to improve the reliability of backend servers. This prevents attackers from using TLS connections to overload backend servers. When a load balancer terminates TLS connections, it acts as a secure gateway for the incoming traffic. Instead of sending encrypted data directly to the backend servers, the load balancer decrypts the data first. This means the backend servers receive unencrypted traffic, which is easier for them to handle because they don't have to spend resources on decryption. As a result, the servers can focus on their primary tasks, such as processing requests and delivering content, without being bogged down by the additional workload of decrypting TLS traffic. This leads to more efficient server performance and reduces the risk of servers being overloaded, especially during high-traffic periods.

- Load balancers can analyze traffic patterns and perform intelligent load-balancing algorithms to distribute requests to servers in a way that maximizes resource utilization.

Overall, health checking, TLS termination, and analytics help improve the reliability and availability of backend servers.

Global and local load balancing

Load balancing is required at both a global and local scale. As shown in *Figure 4.7*, **global server load balancing (GSLB)** distributes traffic across data centers in different geographical regions, while local load balancing focuses on improving resource utilization within a single data center.

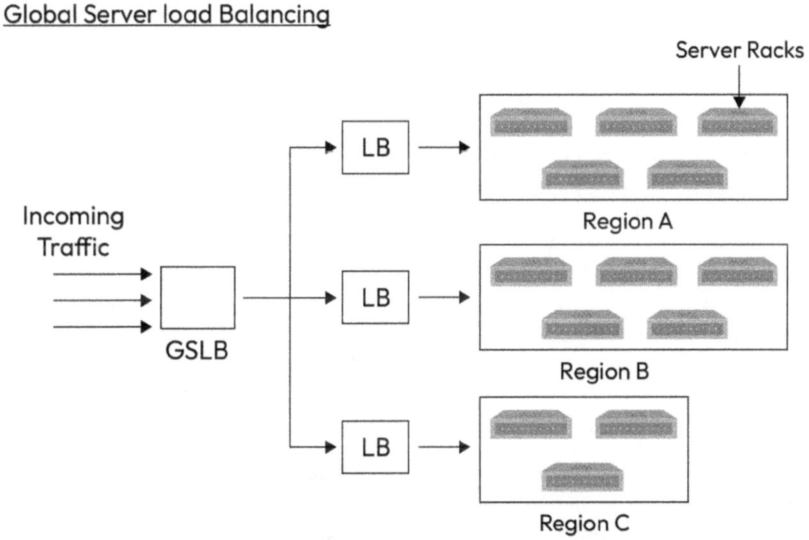

Figure 4.7: Global and local load balancing

GSLB intelligently forwards incoming global traffic to the appropriate data center based on factors such as users' locations, data center health, and the number of hosting servers. It enables automatic failover to alternate data centers during power or network failures. GSLB can be deployed on-premises or obtained as a service (**load balancing as a service (LBaaS)**).

Each data center uses local load balancers to distribute traffic to servers within that region. The local load balancers report health and capacity information to the GSLB, which uses this data to determine how to route incoming traffic to the different data centers.

DNS and GSLB

DNS can also perform GSLB to some extent by returning multiple IP addresses for a DNS query. Different clients will receive those IP addresses in a rotated order, distributing traffic to the different data centers in a round-robin fashion.

However, round-robin DNS load balancing has limitations. It cannot account for uneven demand from different ISPs or detect crashed servers. DNS uses short time-to-live durations for cache entries to enable more effective load balancing.

Overall, both global and local load balancing layers work together to achieve the optimal distribution of incoming traffic across a system with multiple data centers and servers.

While DNS load balancing provides some level of global load distribution, it has limitations that necessitate local load balancers within data centers:

- The small 512-byte size of DNS packets means DNS cannot include all possible server IP addresses in its responses to clients. This limits its load-balancing capabilities.

- Clients can arbitrarily select from the set of IP addresses returned by DNS, potentially choosing addresses for busy data centers. DNS has limited control over client behavior.

- DNS cannot determine the closest server for a given client to connect to, though **geolocation** and **anycasting** techniques could help. However, these are not trivial solutions.

- During failures, recovery through DNS can be slow due to caching, especially when time-to-live values are long.

To address some of these limitations, local load balancers within data centers are needed.

Client applications can connect to local load balancers using virtual IP addresses and these local load balancers distribute the incoming requests from clients to all the available servers.

In summary, while DNS load balancing provides a basic level of global load distribution, local load balancers provide more intelligent and effective load balancing within data centers. They can implement sophisticated techniques to distribute load optimally among backend servers. So, both layers – global and local load balancing – are required to achieve high performance and scalability at both the data center and global levels.

Load balancer algorithms

Load balancers use various algorithms to determine how to distribute incoming client requests to backend servers. Common algorithms include the following:

- **Round-robin scheduling**: This simply assigns each new request to the next server in the list, cycling through the servers in order. This is the simplest algorithm but does not account for differences in server capabilities or load.

- **Weighted round-robin**: This assigns weights to each server based on its capacity. Servers with higher weights receive a proportionally higher number of requests. This accounts for differences in server capabilities.

- **The least connections algorithm**: This assigns new requests to the server with the fewest existing connections. This helps distribute load more evenly among servers, especially when some requests take longer to serve. The load balancer must track the number of connections to each server.

- **The least response time algorithm**: This assigns new requests to the server with the shortest response time, prioritizing performance-sensitive services.

- **IP hash, URL hash, and consistent hashing algorithms**: These assign requests based on hashing the client's IP address or request URL, respectively. This is useful when different clients or URLs need to be routed to specific servers.

In general, load balancers choose algorithms that consider relevant factors such as server capacity and load, request characteristics, and application requirements to achieve their goals of optimizing resource utilization, throughput, response times, and fault tolerance. The simplest algorithms may not distribute the load optimally, while more sophisticated algorithms can require additional state tracking.

Static versus dynamic algorithms

Load balancing algorithms can be classified as either static or dynamic based on whether they consider the state of the servers.

Forwarding decisions are made based on server configuration alone by static algorithms, which are often used in simple load balancing.

Dynamic algorithms do consider the current or recent state of the servers, such as their load levels and health. They maintain state by communicating with the servers, which adds communication overhead and complexity. Dynamic algorithms require load balancers to exchange information to make decisions, making the system more modular. They also only forward requests to active servers.

While dynamic algorithms are more complex than static algorithms, they provide better load balancing results by making more informed decisions based on up-to-date server state. They can adapt to changes over time, unlike static algorithms.

In practice, dynamic algorithms are preferred because the improved load balancing they provide outweighs their additional complexity. However, for simple use cases with few servers, static algorithms may be sufficient and easier to implement. The choice of static versus dynamic algorithms involves trade-offs between simplicity, performance, and adaptability.

Overall, maintaining an accurate state of the backend servers – whether their load levels, response times, or health status – allows load balancers to make the most effective forwarding decisions to optimize resource utilization and meet performance goals.

Stateful versus stateless

Load balancers can be classified as **stateful** or **stateless** based on whether they maintain session information for clients.

Stateful load balancers maintain state information that maps incoming clients to backend servers. They incorporate this state into their load-balancing algorithms to make forwarding decisions. However, stateful load balancers require all load balancers to share state information, which increases complexity and limits scalability.

Stateful load balancers maintain a data structure that tracks the session information for all clients connected to the backend servers. When new requests come in for existing sessions, the load balancer can route them to the correct backend server based on the stored session information.

In comparison, stateless load balancers do not maintain any client session state. They use consistent hashing algorithms to map requests to servers. While this makes them faster and more scalable, they may not be as resilient during infrastructure changes since consistent hashing alone is not enough to route requests to the correct server. Local state is still required along with consistent hashing in some cases.

In general, maintaining state information across multiple load balancers is considered stateful load balancing, while maintaining a local internal state within a single load balancer is stateless load balancing.

The choice between stateful and stateless load balancing involves trade-offs between resilience, scalability, and complexity. Stateful approaches can provide higher availability and reliability, while stateless approaches tend to be faster, more scalable, and simpler to implement. Most production systems employ a hybrid approach.

Overall, both stateful and stateless load balancing techniques are useful depending on the requirements and architecture of the system they are balancing. Let us now discuss the types of load balancers.

Load balancing at different layers of the Open Systems Interconnection (OSI) model

Load balancers can operate at different layers of **the OSI model**, impacting their capabilities and performance. The OSI model is a conceptual framework used to understand and standardize the functions of a computing system without regard to its underlying internal structure and technology. It is divided into seven layers, each specifying particular network functions such as physical transmission, data link, network routing, transport, session management, presentation formatting, and application services. This layered approach helps in the development, troubleshooting, and management of network protocols and communication between different systems and devices, promoting interoperability and flexibility in network architecture.

Layer 4 load balancers distribute load based on the transport layer protocols, TCP and UDP. These load balancers provide stability for connection-oriented protocols since they maintain the TCP/UDP connection and forward requests from a client to the same server on the backend. Moreover, some load balancers help with TLS termination, too. **Layer 7** load balancers distribute the load based on the application layer data from protocols such as HTTP. They can make more intelligent forwarding decisions based on HTTP headers, URLs, cookies, and other application-specific information such as user IDs. In addition to TLS termination, Layer 7 load balancers can perform functions such as rate limiting, HTTP routing, and header rewriting.

While **Layer 7** load balancers can take into account more context from the application to optimize load distribution, **Layer 4** load balancers tend to be faster since they operate at a lower level.

In practice, most load-balancing solutions employ a combination of both Layer 4 and **Layer 7** techniques to balance performance requirements, application needs, and protocol support. **Layer 4** load balancers provide a basic level of load distribution and stability for TCP-based applications, while **Layer 7** load balancers enable more intelligent and customized load distribution for application-specific use cases.

The choice of **Layer 4** versus **Layer 7** load balancing depends on factors such as the application architecture and protocols used, performance requirements, and the level of control and optimization needed over load distribution. Both approaches have benefits for achieving the goals of high availability, scalability, and resource utilization. We will now look at the deployment of load balancers.

Deployment of load balancers

In a typical data center, load balancing usually involves different layers, each with its own role:

- Tier 0 uses the DNS system to switch between multiple IP addresses for a certain website or service.

- Tier 1 uses special routers to split internet traffic based on things such as IP address or simple rules, such as taking turns (round-robin). These routers make it easier to add more load balancers as needed.

- Tier 2 uses **Layer 4** load balancers to make sure that all pieces of data related to a single online activity go to the same next-level load balancer. They use techniques such as consistent hashing and keep track of changes in the network setup.

- Tier 3 uses **Layer 7** load balancers that are in direct touch with the main servers. These balancers check how well the servers are working and share the work between servers that are up and running well. They also take on some tasks to make the servers work more efficiently. Sometimes, these balancers work closely with the main servers themselves.

This layered setup offers benefits such as being able to scale up easily, staying available, being resilient, and making smart use of resources at each layer. The first few layers handle basic traffic splitting, while the higher layers use more information to distribute the load in a smarter way. Together, this makes the entire system work better, faster, and more reliably.

Implementing load balancers

Load balancers can be implemented in various ways to meet the needs of different organizations and applications:

- Hardware load balancers were the first type of load balancers, introduced in the 1990s. They work as standalone devices and can handle many concurrent users due to their performance. However, they are expensive and have some drawbacks. They are difficult to configure, have high maintenance costs, and vendor lock-in issues. While availability is important, additional hardware is needed for failover.

- Software load balancers have become more popular due to their flexibility, programmability, and lower cost. They scale well as requirements grow and availability is easier to achieve by adding additional commodity hardware. They can also provide predictive analytics to prepare for future traffic.

- Cloud load balancers, known as LBaaS, are offered by cloud providers. Users pay based on usage or service-level agreements. They can perform global traffic management between cloud regions in addition to local load balancing. Their advantages include ease of use, scalability, metered costs, and advanced monitoring capabilities.

In general, software and cloud load balancers are gaining popularity due to their benefits over hardware load balancers. They provide a more cost-effective and manageable solution for most organizations today.

However, hardware load balancers still offer the best performance for extremely high throughput requirements. A hybrid approach employing different load balancer types can optimize performance, availability, and cost for complex systems.

The ideal load balancer implementation depends on factors such as the system architecture, throughput needs, management requirements, costs, and resource availability. Each type has pros and cons, so the solution needs to be chosen based on an organization's unique needs and goals. Let us now discuss our last building block, the application gateway/API gateway.

Application gateways

While load balancers provide basic traffic distribution services, application gateways offer more advanced intermediary proxy functionality tailored for modern cloud-based environments.

An application gateway sits between clients and backend services, intercepting traffic to provide routing, security, acceleration, analytics, and adaptability capabilities. Application gateways are especially beneficial for architectures based on microservices, where numerous independent services must be aggregated into unified APIs.

We will cover the key features and benefits of application gateways, their importance for microservices, real-world implementation options, and critical design considerations when integrating them into cloud architectures. *Figure 4.8* shows a typical API gateway, which is the first point of contact for incoming API traffic and covers different aspects of security, AuthN, AuthZ, caching, and even load balancing.

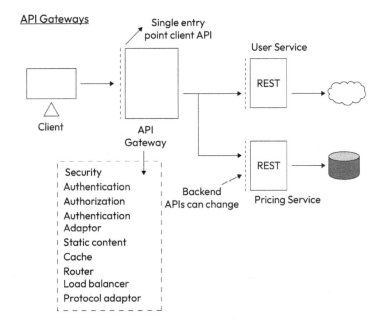

Figure 4.8: A typical API gateway

Let us now explore the features and capabilities of the application gateways (also referred to as API gateways in many cases) in more detail.

Features and capabilities

Application gateways differ from basic load balancers in the more specialized proxy services they provide:

- **Advanced request routing**: Application gateways route incoming requests to the appropriate backend service through rules matching on factors such as hostname, path, headers, and source IP. This advanced routing is crucial in microservices environments comprised of diverse dynamic services that require more logic to map requests.

- **Security**: Gateways centralize shared security services, protecting all backend applications and services. These include authentication, access controls, TLS termination, DDoS (Distributed Denial of Service) protection, and integrating **web application firewalls** (**WAFs**) to guard against OWASP (Open Web Application Security Project) threats.

- **Acceleration and offloading**: To improve performance, gateways implement features such as caching, compression, TCP connection management, and TLS termination so that backend services are not burdened by CPU-intensive tasks. Application gateways absorb these tasks at the edge.

- **Observability**: As a centralized intermediary, gateways aggregate logging, metrics, and traces providing valuable application visibility. Unified insights into all traffic facilitate monitoring, analytics, and debugging.

- **Adaptability**: Gateways enable adapting requests and responses to gracefully handle changing backend service capabilities. For example, protocol translation and request shaping for legacy services.

These capabilities make application gateways well suited for the demands of organizing logic, security, and reliability in modern service-oriented architectures. Many organizations such as Netflix are known for their microservices-based architectures. There are thousands of microservices deployed in production at Netflix. API gateways help with these microservices as well, as we will see in the next section.

Microservices architectures

Microservices approaches decompose monoliths into collections of independent, reusable services. These loosely coupled services communicate through well-defined interfaces using lightweight protocols such as REST/HTTP.

Microservices environments involve numerous discrete services developed and deployed independently. This dynamic system presents challenges in exposing cohesive APIs to clients, handling cross-cutting concerns, and coordinating services.

Application gateways are crucial for microservices to do the following:

- Aggregate numerous microservices into unified logical APIs exposed to clients. This decouples clients from volatile backend implementations.

- Implement service discovery and dynamic request routing to backend services, load balancing across instances.

- Handle cross-cutting concerns such as security, monitoring, and reliability in a centralized place, avoiding code duplication across services.

- Enable releasing updated gateway versions independently of backend services.

- Accelerate development by allowing services to be developed and scaled independently.

In short, application gateways provide the abstraction layer to coordinate microservices into cohesive systems and handle common integration needs. Different cloud providers have different approaches to application gateway implementation, and we will briefly talk about some of them in the next section.

Cloud-native implementations

Leading cloud providers offer fully managed application gateway services that integrate tightly with their technology stacks:

- **AWS**: Amazon API Gateway handles API creation, deployment, management, and security. Application Load Balancer routes traffic to AWS services and containers.

- **Microsoft Azure**: Azure Application Gateway provides advanced Layer 7 load balancing along with WAF, SSL (Secure Sockets Layer) offload, end-to-end TLS encryption, and autoscaling capabilities.

- **Google Cloud Platform**: Cloud Armor provides DDoS protection and a WAF that integrates with Google's network infrastructure. Cloud CDN (Content Distribution Network) offers caching and acceleration.

- **Kubernetes environments**: Ingress controllers such as Istio, Kong, Traefik, and Ambassador act as API gateways for Kubernetes clusters, handling ingress HTTP/HTTPS traffic.

These serverless, auto-scaling gateways simplify operational overhead for organizations leveraging cloud-native architectures. Along with cloud-native implementations, many enterprises also choose on-premise options.

On-premises options

Application gateways or API gateways are not only present in cloud environments. For many enterprises, especially in the financial domain, a lot of services and applications are hosted in their own data centers. These enterprises prefer leveraging open-source software that can be seamlessly deployed in on-premise environments. Some popular application gateway platforms for on-premises environments are outlined in the following list:

- **Kong**: Kong Gateway and Kong Mesh offer API gateway and service mesh capabilities using lightweight proxy servers. Plugins support authentication, security, analytics, and more.

- **Tyk**: Tyk API Gateway provides an open-source gateway with robust access control, developer portal capabilities, and REST API-driven configuration.

- **NGINX**: NGINX can be configured as an API gateway and load balancer, with rate limiting, access control, and service discovery integrations.

- **HAProxy**: HAProxy is a fast, lightweight load balancer and proxy that can be configured for Layer 7 application delivery.

These open-source options run on commodity infrastructure, providing flexibility for on-premises gateway implementations. With their advanced routing, security, acceleration, and coordination capabilities, application gateways are integral components for implementing reliable, scalable microservices architectures.

Summary

This chapter provided a comprehensive overview of three fundamental building blocks used in distributed system design – DNS, load balancers, and application gateways. We learned how DNS provides a directory service translating domain names to IP addresses through a globally distributed hierarchy of name servers. Load balancers distribute requests across backend servers using algorithms that optimize performance and reliability. Application gateways act as specialized proxies providing advanced routing, security, acceleration, and coordination logic tailored for modern cloud architectures, especially critical for microservices. In this chapter, we also covered some nuances of these basic building blocks, for example, the DNS caching for lower latency and traffic, some load balancer algorithms and implementations in different layers of the OSI model, and application gateway support for providing a unified front to distributed services on the backend. These are the basic building blocks of modern enterprise systems deployed at scale. Almost every system design interview would require you to describe these systems and how they work.

In the next chapter, we will study another cornerstone of any modern system – **databases** and **storage**.

5

Design and Implementation of System Components –Databases and Storage

In our rapidly evolving digital landscape, where information flows ceaselessly and the demand for data-driven decision-making is unrelenting, the role of databases and storage has never been more pivotal. As we navigate the intricate web of data that defines our modern world, the ability to efficiently collect, store, retrieve, and manage information is paramount. This chapter delves into the very heart of this technological foundation, exploring the fundamental concepts, strategies, and technologies that underpin the organization and preservation of data. We will look into the details of how some of the popular databases and storage systems are designed.

In a world where data is often hailed as the new currency, databases serve as the repositories of knowledge, housing vast troves of information that fuel businesses, drive research, and empower innovation. Yet, without an equally robust and adaptable storage infrastructure, the potential of these databases remains untapped. Together, databases and storage systems form an inseparable duo that enables us to harness the power of data and transform it into actionable insights.

We will cover the following concepts in detail in this chapter.

- Databases
- Key-value stores
- DynamoDB
- Column-family databases
- HBase
- Graph-based databases
- Neo4j

Databases

Databases provide a way to store, organize, and manipulate large amounts of information in a systematic and controlled manner. This data can be structured, semi-structured, or unstructured, depending on the specific type of database and its use case. Let's think about the reasons why we would need a database.

Here are some key reasons why we need databases:

- **Data organization**: Databases provide a structured and organized way to store data. Data is arranged in tables, rows, and columns, making it easy to categorize and access information.

- **Data retrieval**: Databases enable fast and efficient data retrieval. Users can perform complex queries to extract specific data, and indexing mechanisms speed up data lookup.

- **Data integrity**: Databases enforce data integrity by using constraints, relationships, and validation rules. This ensures that data is accurate, consistent, and reliable.

- **Data security**: Databases offer security features such as user authentication, authorization, and encryption to protect data from unauthorized access and breaches.

- **Data consistency**: Through transaction management and **ACID** (**Atomicity, Consistency, Isolation, and Durability**) properties, databases ensure that data remains consistent even when multiple users access and modify it simultaneously.

- **Scalability**: Databases can handle large volumes of data and scale as data needs grow. Both vertical scaling (adding more resources to a single server) and horizontal scaling (adding more servers or nodes) are possible.

- **Redundancy and backup and recovery**: Databases often incorporate redundancy and backup mechanisms to ensure data availability and reliability. This includes features such as replication and automated backups. Databases offer mechanisms to create backups and restore data if there is data loss or a system failure.

- **Complex queries**: Databases allow users to perform complex queries and aggregations, making it possible to extract valuable insights and reports from large datasets.

- **Data relationships**: In relational databases, data relationships are defined, enabling the efficient management of interconnected data, such as customer orders, products, and inventory.

- **Data history**: Some databases maintain historical data, providing a historical view of data changes over time. This is valuable for auditing and compliance purposes.

- **Data analysis**: Databases support data analysis and reporting tools, allowing organizations to make data-driven decisions and gain insights into their operations and performance.

- **Data sharing**: Databases support concurrent access and the sharing of data among different users and applications, making them essential for collaborative and real-time environments.

In essence, databases are the foundation for data-driven applications and systems, from websites and mobile apps to **ERP (enterprise resource planning)** systems and scientific research. They ensure that data is stored, managed, and utilized effectively, facilitating informed decision-making and efficient data processing.

It's important to acknowledge that one kind of database may not be able to solve the needs of different use cases for different systems and applications. Let's learn about the various types of databases.

Types of databases

Databases can be broadly classified as **relational** and **NoSQL** databases.

In both cases, databases aim to provide efficient data storage, retrieval, and management, but they do so using different models and approaches. The choice between these two types of databases depends on the specific needs of an application or system, with factors such as data structure, scalability, and performance considerations playing a significant role in the decision.

Now, let's take a look at these two database types in more detail.

Relational databases

Relational databases are a type of database that organizes and store data in a tabular format, consisting of rows and columns. They are based on the principles of the relational model, which was introduced by **Edgar F. Codd** in the 1970s. This model defines relationships between data elements and enables efficient data storage and retrieval.

Key characteristics and concepts of relational databases include the following:

- **Tables**: In a relational database, data is organized into tables. Each table represents an entity or concept, such as "customers," "products," or "orders." Tables are further divided into rows and columns.

- **Rows**: Each row in a table, often referred to as a "record" or "tuple," represents a unique data entry. For example, in a "customers" table, each row corresponds to an individual customer, with each column containing specific information about that customer, such as their name, address, and phone number.

- **Columns**: Columns, also known as "attributes" or "fields," define the type of data that can be stored in a table. Each column has a name and a data type, such as text, numeric, date, or binary.

- **Keys**: Relational databases use keys to establish relationships between tables. The primary key uniquely identifies each row in a table, while foreign keys in one table refer to the primary key in another table, creating relationships between them.

- **Normalization**: The process of normalization is used to eliminate data redundancy and improve data integrity. It involves breaking down tables into smaller, related tables to reduce duplication and maintain consistency.

- **Structured Query Language (SQL):** SQL is the language used to interact with relational databases. It provides a standardized way to create, retrieve, update, and delete data, as well as to define the structure of tables and establish relationships between them.

- **ACID properties**: Relational databases are known for their strong support of ACID properties, which ensure data consistency, reliability, and durability. ACID is crucial for maintaining data integrity.

- **Transactions**: Relational databases enable transactions, which are sequences of one or more SQL operations that are executed as a single unit. If any part of a transaction fails, the entire transaction can be rolled back, ensuring data consistency.

Some well-known **relational database management** systems (**RDBMS**) include the following:

- MySQL
- PostgreSQL
- Oracle Database
- Microsoft SQL Server
- SQLite
- IBM Db2

Relational databases are widely used in various applications and industries where data consistency, structure, and reliability are critical, such as financial systems, inventory management, **customer relationship management** (**CRM**), and many others. However, it's essential to choose the right RDBMS and database design to match the specific needs of an application.

There is another class of databases that is very different from the relational database. We will learn about NoSQL databases in this next section.

NoSQL databases

Non-relational databases, often referred to as **NoSQL** databases (which stands for **Not Only SQL**), are a category of database management systems that depart from the traditional relational database model. Unlike relational databases that use tables, rows, and columns, NoSQL databases offer more flexibility in data storage and retrieval. These databases are designed to handle unstructured or semi-structured data, making them suitable for various types of applications and use cases.

Here are some types of NoSQL databases:

- **Key-value stores:** Key-value stores associate a unique key with a data value. They are highly efficient for simple data retrieval but are less suitable for complex queries. Examples include Redis, Amazon DynamoDB, and Riak.

- **Document-oriented databases**: These databases store data as documents, such as JSON or XML, and are commonly used in web applications. Examples include MongoDB, CouchDB, and RavenDB.

- **Column-family databases**: Column-family stores organize data into column families, which are groups of related data. These databases are known for their ability to scale horizontally. Examples include Apache Cassandra, HBase, and ScyllaDB.

- **Graph-based databases**: Graph databases are designed for data with complex relationships. They use graph structures to represent and navigate relationships between data points. Examples include Neo4j, Amazon Neptune, and OrientDB.

The key benefits of NoSQL databases include their ability to handle large volumes of data, support scalability, and adapt to evolving data structures and requirements. However, the trade-off is typically a decreased emphasis on strong data consistency and the need for more careful consideration of data modeling and indexing.

NoSQL databases are used in a variety of applications, such as content management systems, real-time analytics, **IoT** (**Internet of Things**), and social media platforms. The choice of a NoSQL database type depends on the specific data needs and characteristics of an application.

The advantages and disadvantages of relational and NoSQL databases

Now that we have learned about the two types of databases, let's summarize the advantages and disadvantages of both these types.

Relational databases

Here are the advantages of relational databases:

- **Reduced data redundancy**: Relational databases enforce data integrity by minimizing data duplication across tables. This optimizes storage space and simplifies data retrieval.

- **Robust security features**: Built-in security measures safeguard sensitive information from unauthorized access.

- **ACID transactions**: Native support for ACID ensures data consistency and reliability during operations.

Here are the disadvantages of relational databases:

- **Performance**: Complex joins and fetching data from multiple tables can lead to slower performance.

- **Memory-intensive**: Rows and columns consume storage space, even for `null` values, increasing memory requirements.

- **Complexity**: Managing intricate joins and complex relationships can add significant complexity to database administration.

A lack of horizontal scalability: Traditional relational databases are difficult to scale horizontally, making them unsuitable for large data volumes.

NoSQL databases

Here are the advantages of NoSQL databases:

- **A flexible data model**: NoSQL databases excel in storing both structured and unstructured data types. This adaptability allows for diverse data formats and easy accommodation of changing data requirements.

- **Schema updates**: Schema updates in NoSQL databases can be done without disrupting applications, facilitating an evolving data model.

- **Horizontal scaling**: Superior horizontal scaling capabilities enable seamless expansion of database capacity to handle growing datasets and workloads.

Here are the disadvantages of NoSQL databases:

- **A lack of standardization**: The absence of standardization leads to a wider range of designs and query languages compared to relational databases. This can create challenges in interoperability and data consistency across different NoSQL implementations.

- **Limited ACID transactions**: Most NoSQL databases lack support for ACID transactions (except for some specialized types). This might impact the reliability and integrity of data operations that require strict consistency guarantees.

By understanding these strengths and weaknesses, you can effectively choose the database type that best aligns with your project's specific requirements and performance needs.

In the following sections, we will learn about various types of databases in more detail, such as key-value stores, **DynamoDB**, **HBase** (a column-oriented database), and **Neo4j** (a graph database).

Key-value stores

In any sophisticated software system, data storage and retrieval is a fundamental concern. **Key-value stores** provide a simple, efficient, and highly scalable solution to store data. This chapter delves into the intricacies of designing a robust key-value store while emphasizing key concepts such as scalability, replication, versioning, configurability, fault tolerance, and failure detection.

This chapter will equip you with the knowledge to design and manage a resilient key-value store. We'll embark on this journey by initially defining the requirements of a key-value store and designing its API. Then, we'll explore techniques to ensure scalability, using consistent hashing and strategies to replicate partitioned data. Furthermore, we'll uncover how to manage versioning and resolve conflicts that arise due to concurrent modifications. Lastly, we'll dive into how to make the key-value store fault-tolerant and devise mechanisms for timely failure detection. Let's begin by understanding what a key-value store is.

What is a key-value store?

A key-value store, also known as a key-value database, is a simple data storage paradigm where each unique key corresponds to a particular value. Think of it as a large, distributed dictionary or a **Distributed Hash Table (DHT)**, where data can be stored, retrieved, and updated using an associated key. This key-value pair forms the fundamental unit of data storage, and it's this simplicity in design that aids efficient read and write operations. Let's now look at why it is useful in distributed systems.

Use in distributed systems

In the realm of distributed systems, key-value stores play a pivotal role due to their high performance, scalability, and ease of use. Here are a few reasons why:

- **Scalability**: Given the simplicity of key-value pairs, these stores can easily distribute data across multiple nodes, thereby improving a system's capacity and throughput.

- **Performance**: Key-value stores typically offer quick read and write access, especially if the key-value pair resides on the same node.

- **Flexibility**: Unlike relational databases, key-value stores do not require a fixed data schema. This allows for the storage of structured, semi-structured, or unstructured data.

- **Fault-tolerance**: Key-value stores can be designed to replicate and partition data across multiple nodes. This ensures that data is still accessible if there is a node failure.

Hence, key-value stores provide an efficient and flexible approach to data storage and retrieval in distributed systems, making them an integral part of modern software architecture. Let's now learn about some of the functional and non-functional requirements of designing a key-value store.

Designing a key-value store

When embarking on the design of a DHT or key-value store, we must first establish its functional and non-functional requirements.

Functional requirements

The following are some of the functional requirements:

- **Put(key, value)**: A system should support a `put` operation that inserts a key-value pair into the store. If the key already exists, the corresponding value should be updated.

- **Get(key)**: A system should support a `get` operation that retrieves the value associated with a specified key. If the key does not exist, the operation should return an appropriate error message.

- **Delete(key)**: A system should support a `delete` operation that removes a given key-value pair from the store. If the key does not exist, the operation should return an appropriate error message.

Non-functional requirements

The following are some of the non-functional requirements:

- **Scalability**: As key-value stores are often used in high-demand scenarios, a system must be able to scale horizontally (add more nodes to the system) to serve a growing amount of data and traffic.

- **Performance**: A system should ensure low latency for `put` and `get` operations. Even as data grows and spans multiple nodes, the time taken for these operations should not degrade significantly.

- **Durability**: Once a value is stored in a system, it should persist. The system should ensure data is not lost due to node failures.

- **Consistency**: A system should ensure that all read operations reflect the most recent write operation for a given key. If writes occur simultaneously, this might involve resolving conflicts.

- **Availability**: A system should remain available for operations despite node failures. This would require replication of data across multiple nodes.

- **Partition tolerance**: The system should function and maintain data integrity even when network failures occur and nodes are unable to communicate.

Having understood the functional and non-functional requirements of a key-value store, let's now delve into the specifics of ensuring scalability and replication. This next section will provide insight on how to accommodate growing data and traffic, as well as how to achieve data replication across multiple nodes for higher availability.

Enhancing scalability and data replication

In this section, we will explore how consistent hashing can bolster scalability and how to replicate partitioned data efficiently.

Boosting scalability

One of the essential design requirements for our system is scalability. We store key-value data across multiple storage nodes. Depending on demand, we might need to augment or diminish these storage nodes. This implies that we must distribute data across all nodes in the system to evenly distribute the load.

For instance, consider a scenario where we have four nodes, and we aim to balance the load equally by directing 25% of requests to each node. Traditionally, we would use the modulus operator to achieve this. Each incoming request comes with an associated key. On receiving a request, we calculate the hash of the key and then find the remainder when the hashed value is divided by the number of nodes (m). The remainder value (x) indicates the node number to which we route the request for processing. *Figure 5.1* shows a key that is hashed, and a modulo operation is applied to the result to determine the node to which the request carrying that key-value pair should be routed.

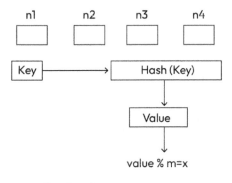

Figure 5.1: A modulo-based key-value pair routing

However, this method falls short when we add or remove nodes, as we end up having to move a significant number of keys, which is inefficient. For example, if we remove node 2, the new server to process a request will be node 1 because 10%3 equals 1. Given that nodes maintain information in local caches, such as keys and their values, we need to transfer this data to the next node tasked with processing the request. However, this can be costly and result in high latency.

Next, let's delve into efficient data copying methods.

Consistent hashing

We covered consistent hashing in *Chapter 3*. Let's refresh what we learned here. Consistent hashing offers a powerful way to distribute load across a set of nodes. In this approach, we assume a conceptual ring of hashes, ranging from 0 to *n-1*, where *n* represents the total number of available hash values. We calculate the hash for each node's ID and map it onto the ring. The same process is applied to requests. Each request is completed by the next node found when moving in a clockwise direction on the ring. *Figure 5.2* shows an example of the consistent hashing scheme, where a request carries a key-value pair, and the key is hashed to a result. The result of the hash is then mapped to a location on the ring, and the request is sent to the next node – **N3**, in this case.

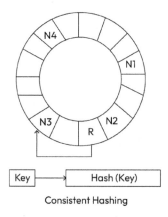

Consistent Hashing

Figure 5.2: Consistent hashing of key-value pairs in requests

When adding a new node to the ring, only the immediate next node is affected, as it shares its data with the newly added node. Other nodes remain unaffected. This allows us to scale easily, as we keep changes to our nodes minimal, with only a small portion of overall keys needing to move. As the hashes are randomly distributed, we expect the request load to be randomly and evenly distributed on average on the ring.

However, consistent hashing doesn't always ensure an equal division of the request load. A server handling a large chunk of data can become a bottleneck in a distributed system, reducing the overall system performance. These are referred to as hotspots.

Virtual nodes

To achieve a more evenly distributed load across the nodes, we can utilize virtual nodes. Instead of applying a single hash function, we apply multiple hash functions to the same key.

For instance, if we have three hash functions, we calculate three hashes for each node and place them onto the ring. For the request, we use only one hash function. Wherever the request lands on the ring, it's processed by the next node found when moving in a clockwise direction. Each server has three positions, so the request load is more uniform. Furthermore, if a node has more hardware capacity than others, we can add more virtual nodes by using additional hash functions. This way, it'll have more positions in the ring and serve more requests.

The advantages of virtual nodes

Virtual nodes offer the following benefits:

- If a node fails or undergoes routine maintenance, the workload is uniformly distributed over other nodes. For each newly accessible node, the other nodes receive nearly equal load when it comes back online or is added to a system.

- Each node can decide how many virtual nodes it's responsible for, considering the heterogeneity of the physical infrastructure. For example, if a node has roughly double the computational capacity compared to the others, it can handle more load.

Now that we've made our key-value storage design scalable, the next step is to make our system highly available. To ensure high availability, we need to introduce replication strategies and mechanisms to handle failures, which will be the focus of the following sections.

Data duplication strategies

There are several ways to duplicate data in a storage system. The two main methods are the **primary-secondary model** and the **peer-to-peer model**.

The primary-secondary model

In this model, one storage area is designated as the primary, while the others act as secondary storage areas. The primary storage area handles write requests, while the secondary storage areas handle read requests and duplicate their data from the primary storage area. However, there is a delay in replication after writing. If the primary storage fails, the system loses its write capability, making it a single point of failure. *Figure 5.3* shows an example of this model, where writes to the primary storage are replicated to a secondary storage system, and the reads can be served by additional read-only replicas as well.

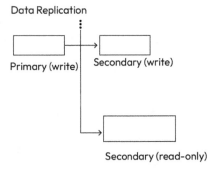

Figure 5.3: A primary-secondary data replication model

The peer-to-peer model

In contrast, the peer-to-peer model designates all storage areas as primary. They can all handle both read and write requests and replicate data among themselves to stay up to date. However, duplicating data across all nodes is often inefficient and expensive. A common solution is to select a smaller number, such as three to five nodes, for replication. *Figure 5.4* shows a peer-to-peer data replication model in which all nodes have all the writes persisted and are used to service reads.

Peer to Peer Model

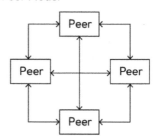

Figure 5.4: A peer-to-peer data replication model

We'll use the peer-to-peer model for our data duplication because of the latency and availability advantage it provides. With the primary-secondary model, we have a single point of failure if the primary is unavailable; we can avoid that completely by using the peer-to-peer model for data duplication. This model will help us achieve durability and high availability by replicating data on multiple hosts. We'll replicate each data item on n hosts, where n is a parameter configured for each instance of the key-value store. For instance, if we set n to five, our data will be replicated across five nodes.

Each node will copy its data to other nodes. The node responsible for handling read or write operations is called the coordinator. The coordinator node is directly responsible for specific keys. For example, if a coordinator node is assigned the key K, it is also responsible for duplicating these keys to n-1 successors on the ring (going clockwise). These lists of successor virtual nodes are known as preference lists. To prevent placing replicas on identical physical nodes, the preference list can bypass those virtual nodes whose physical node is already listed. Let's now discuss some of the nuances in implementing the get and the put functions in our key-value store.

Implementing get and put functions

This section will delve into how to implement get and put functions in our key-value store.

Implementing get and put operations

A key requirement for our system is configurability. This means the ability to adjust the balance between availability, consistency, cost-effectiveness, and efficiency. We can achieve this by incorporating the fundamental get and put functions of a key-value store.

In our system, every node can perform `get` (read) and `put` (write) operations. The node that manages these operations is known as a coordinator, which is usually the first among the top n nodes on the preference list.

Clients can select a node in two ways:

- By routing the request through a generic load balancer
- By using a partition-aware client library that directs requests straight to the relevant coordinator nodes

Both methods have their advantages. The first approach doesn't tie the client to the code, while the second one can achieve lower latency due to a reduced number of hops, as the client can directly access a particular server.

We can make our service configurable by controlling the balance between availability, consistency, cost-effectiveness, and performance. One way to do this is by using a protocol similar to those used in quorum systems.

Let's assume n from the top n of the preference list equals 3. This implies that three copies of the data need to be maintained. If nodes are placed in a ring and A, B, C, D, and E are the nodes in clockwise order, then if a write operation is performed on node A, the data copies will be placed on nodes B and C. These are the next nodes found when moving in a clockwise direction on the ring.

Using r and w

Consider two variables, r and w. The former represents the minimum number of nodes required for a successful read operation, while the latter signifies the minimum number of nodes involved in a successful write operation. Therefore, if r equals 2, our system will read from two nodes when data is stored across three nodes. We need to select values for r and w such that at least one node is common between them. This ensures that readers can access the latest-written value. To achieve this, we'll use a quorum-like system by setting $r + w > n$.

- Here's an overview of how the values of n, r, and w impact the speed of reads and writes:

n	r	w	Description
3	2	1	Violates constraint – $r + w > n$
3	2	2	Fulfills constraint
3	3	1	Slow reads and fast writes
3	1	3	Slow writes and fast reads

Table 5.1: The impact of selecting different numbers of read and write successful requests on our system

If we design our system such that more nodes have to return success on reads for an incoming read request from the client to succeed, we will have slow reads, and similar results for the write requests as well. The right balance is to design quorums such that we do not compromise reads or writes but have enough redundancy to support arbitrary node failures.

If n equals 3, which means we have three nodes where the data is copied, and w equals 2, the operation ensures that writing to two nodes makes this request successful. The third node updates the data asynchronously.

In this model, the latency of a `get` operation is determined by the slowest of the r replicas. This is because a larger r value prioritizes availability over consistency.

We've now met the requirements for scalability, availability, conflict resolution, and a configurable service. The final requirement is to have a fault-tolerant system, which we'll discuss in the next lesson.

Ensuring fault tolerance and identifying failures in a key-value store

In this section, we will explore how to construct a fault-tolerant key-value store capable of identifying and managing system failures. Let's begin with the techniques to manage temporary failures, ensuring that our key-value store can weather short-term disturbances or disruptions.

Managing temporary failures

A common approach to dealing with failures in distributed systems is the use of a **quorum-based system**. A quorum refers to the minimum number of votes that a distributed transaction needs to carry out an operation. If a server that is part of the consensus goes down, the operation cannot proceed, impacting a system's availability and durability.

Instead of relying on strict quorum membership, we propose using a sloppy quorum. In most cases, a central leader coordinates communication between consensus participants. After a successful write, participants send an acknowledgment. The leader responds to the client upon receipt of these acknowledgments. However, this system is vulnerable to network outages. If the leader temporarily goes down and the participants cannot reach it, they declare the leader dead. This necessitates the election of a new leader. Frequent elections can hamper performance, as the system spends more time choosing a leader than performing actual tasks.

In a sloppy quorum, the first n healthy nodes from the preference list handle all read and write operations. These n healthy nodes may not be the initial n nodes identified when moving clockwise in the consistent hash ring.

Consider a configuration where n equals 3. If node A is temporarily unavailable or unreachable during a write operation, the request is sent to the next healthy node from the preference list, which in this case is node D. This ensures the required availability and durability. After processing the request, node D includes a hint about the intended recipient node (in this case, A). Once node A is back online, node D sends the request information to A to update its data. After the transfer is complete, D removes this item from its local storage, keeping the total number of replicas in the system unchanged.

Figure 5.5 shows the flow of data in this case, where initially a write request is forwarded to node A, which may be unavailable at the time, and so the request ends up in node D. This node persists the request information along with the fact that the request was originally intended for A. When node A is available again, node D forwards all the requests that it served, which were originally intended for node A, and cleans up its internal state.

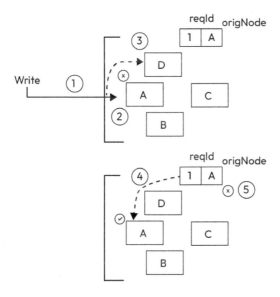

Figure 5.5: A hinted handoff

This approach, known as a hinted handoff, guarantees that read and write operations are fulfilled even if there is temporary node failure.

Addressing permanent failures

In the face of permanent node failures, it's crucial to maintain synchronized replicas for enhanced system durability. The goal is to quickly detect discrepancies among replicas and minimize data transfer. To this end, we employ **Merkle trees**.

A Merkle tree hashes individual key values and uses these hashes as tree leaves. Parent nodes higher up in the tree contain hashes of their child nodes. Each branch of the Merkle tree can be independently verified, negating the need to download the entire tree or dataset. Merkle trees reduce the volume of data transferred during inconsistency checks across copies. Synchronization isn't necessary if, for instance, two trees' root hashes and leaf nodes are identical. Hosts can identify out-of-sync keys as they exchange the hash values of children, continuing until they reach the tree leaves. This anti-entropy mechanism ensures data consistency while reducing data transmission and disk access during synchronization.

Here's how Merkle trees function:

- Hash all keys to create leaf nodes.

- Each node maintains a unique Merkle tree for the key range it hosts for every virtual node. Nodes can verify the correctness of keys within a given range. Two nodes exchange the Merkle tree root corresponding to common key ranges. The comparison proceeds as follows:

    I.   Compare the root node hashes of Merkle trees.

    II.   If they're identical, don't proceed.

    III. Using recursion, traverse the left and right children. Nodes identify any differences and synchronize accordingly.

The advantage of using Merkle trees lies in their ability to independently verify each branch without downloading the entire tree or dataset. This reduces the volume of data exchanged during synchronization and the number of disk accesses required during the anti-entropy process.

However, the downside is that when a node joins or leaves a system, tree hashes must be recalculated, as multiple key ranges are affected.

To ensure other nodes in the ring detect a node failure, we need to incorporate this into our design.

Ring membership promotion for failure detection

Nodes may be offline briefly or indefinitely. We shouldn't rebalance partition assignments or repair unreachable replicas when a single node goes down, as departures are rarely permanent. Therefore, adding and removing nodes from a ring should be done cautiously.

Planned commissioning and decommissioning of nodes lead to membership changes. These changes form a history, recorded persistently on each node's storage and reconciled among ring members using a gossip protocol. This protocol also maintains an eventually consistent view of membership. When two nodes randomly select each other as peers, they can efficiently synchronize their persisted membership histories.

Here's how a gossip-based protocol works. Suppose node A starts up for the first time and randomly adds nodes B and E to its token set. The token set has virtual nodes in the consistent hash space and maps nodes to their respective token sets. This information is stored locally on the disk space of the node.

Now, suppose **node A** handles a request, resulting in a change. It communicates this to **nodes B** and E. Another node, **D**, has **nodes C** and E in its token set. It makes a change and informs **nodes C** and E. The other nodes follow the same process. Eventually, every node becomes aware of every other node's information. This method efficiently shares information asynchronously without consuming much bandwidth.

A key-value store offers flexibility and scalability for applications dealing with unstructured data. Web applications can use key-value stores to store information about a user's session and preferences. By using a user key, all data becomes accessible, making key-value stores ideal for quick read and write operations. Key-value stores can power real-time recommendations and advertising, as they can quickly access and present fresh recommendations.

A system design interview – key value store design questions and strategies

During a system design interview, an interviewer may ask you to design a key-value store. This request tests your understanding of the basic principles of distributed systems, your ability to think about scale, and your familiarity with trade-offs in consistency and availability.

When designing a key-value store, remember these key strategies:

- **Clearly define the problem**: Start by understanding the requirements and constraints of the problem. A key-value store could be asked for in many contexts – for a small application or a large-scale system – so clarify the necessary parameters first.

- **Focus on scalability and performance**: Discuss how you would ensure the key-value store can handle increasing amounts of data and requests. You might discuss sharding the data and using consistent hashing to distribute keys.

- **Discuss replication and consistency**: Address how you would handle data replication to achieve high availability. Also, discuss consistency models and how you'd handle write conflicts in replicated data.

- **Plan for failure**: Remember that distributed systems can and will fail. Discuss strategies to achieve fault tolerance, such as using redundancy and having a strategy for failure detection and recovery.

- **Think about evolvability**: Your system should be able to evolve over time. Discuss how you might handle data versioning and how you'd make the system configurable to meet changing needs.

Remember, in a system design interview, your ability to communicate your thought process and justify your design decisions is often as important as getting the "right" answer.

DynamoDB

DynamoDB is a fully managed, serverless NoSQL database service provided by **Amazon Web Services** **(AWS)**. It provides fast and predictable performance with seamless scalability. DynamoDB is a key-value and document database that uses SSD storage and is spread across three geographically distinct data centers. It is highly available, with replication across multiple availability zones. DynamoDB is a great choice for applications that need very low latency access to data, the ability to scale storage and throughput up or down as needed, and high availability and durability of data.

Let's understand some aspects of the DynamoDB design that are useful as system design practitioners. Some of our design principles of a generic key-value store are directly applicable to DynamoDB.

No fixed schema

DynamoDB aims to have no fixed schema in its design. This allows DynamoDB to support a multitude of applications and use cases. To meet the diverse needs of a broad customer base, a database design must be versatile, scalable, and high-performing, which is fulfilled by a NoSQL database.

Unlike RDBMS, which requires a predetermined schema to build indexes, NoSQL databases offer flexibility by deferring schema decisions to read time. This enables easy API integration and accommodates various use cases.

The benefits of NoSQL

- **Flexibility**: NoSQL databases can store unstructured or semi-structured data, allowing data from multiple tables in a normalized RDBMS to reside in a single document. This ease of use simplifies API coding and enhances functionality.

- **Scalability**: NoSQL databases store data in documents rather than tables, simplifying scaling processes. Unlike RDBMS, which is tightly linked to its storage hardware, NoSQL can easily distribute its databases across large clusters, providing a more straightforward scaling mechanism.

- **Performance**: The data models used in NoSQL databases are engineered for optimal performance, which is especially important for large-scale operations.

- **Availability**: NoSQL databases ensure high availability by enabling seamless node replacement and easier partitioning. This feature also minimizes downtime during node failures by rerouting requests to replica shards.

In the NoSQL setup, data is organized into tables built atop a key-value store. Tables may contain zero or more items identified by primary keys. Each item consists of one or more attributes, considered as basic data types such as strings or numbers.

API functions

The following are some of the API functions:

- `PutItem`: Adds or replaces an item based on the input key

- `UpdateItem`: Modifies an existing item or creates a new one if it doesn't exist

- `DeleteItem`: Removes an item identified by its primary key

- `GetItem`: Retrieves an item's attributes based on its primary key

Partitioning data in DynamoDB

In DynamoDB, data is partitioned horizontally across multiple storage servers. To recap, there are two ways to partition data – vertically or horizontally. For vertical partitioning, there is a need to know the schema beforehand. Since DynamoDB has no schema, and since we need to support a large number of rows, horizontal partitioning is the preferred option. Each table will be split into partitions, with each partition backed by SSD storage.

Figure 5.6 shows the vertical and horizontal partitioning of data.

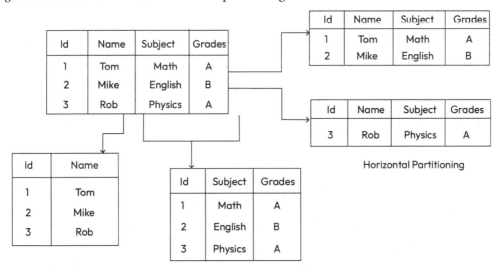

Figure 5.6: Vertical and horizontal data partitioning in DynamoDB

Primary key types

To locate an item in a partition, for a lookup or update, there are two schemas – a partition key, and a partition key with a sort key (also called a composite key):

- **A partition Key**: Determines an item's storage location through a hash function.
- **A composite key**: Consists of a partition key and a sort key. The hash function output, coupled with the sort key, identifies the item's storage location.

Figure 5.7 shows an example of the partition key, sort key, and composite keys in DynamoDB. The hash is applied to the partition key, which along with the sort key, determines the location of the key on the backend storage node.

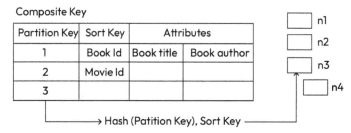

Figure 5.7: Partition keys, sort keys, and composite keys in DynamoDB

Secondary indexes

DynamoDB's design accommodates alternative querying keys, in addition to primary keys, providing more query options.

Throughput optimizations in DynamoDB

In the context of database management, specifically in DynamoDB, the optimization of throughput allocation is of utmost importance. Throughput, in this case, refers to the rate at which a system can fulfill read or write requests. Efficiently partitioning the tables and managing throughput can lead to increased performance and reduced downtime.

In the next section, we will briefly cover read and write capacity units in DynamoDB and ways in which bursting and adaptive capacity management are used to increase throughput.

Throughput allocation

DynamoDB allows you to set a provisioned throughput, the upper limit of **read capacity units** and **write capacity units** (**RCUs** and **WCUs**) that a system will allocate for your tables. Initial partitioning spreads this allocated throughput equally across all partitions, assuming each key within those partitions will be accessed uniformly. However, this is often not the case in real-world applications, leading to inefficiencies such as underutilized or overloaded partitions.

RCUs and WCUs are metrics that gauge a system's ability to complete read and write requests, respectively. These units are crucial when discussing throughput optimization. For example, if a table has a provisioned throughput of 20,000 RCUs and 5,000 WCUs, it implies that the system can, at maximum capacity, read 20,000 items and write 5,000 items per second for an item of a given arbitrary size.

In DynamoDB, you may need to add or remove partitions based on data storage or throughput needs. When you alter the number of partitions, the provisioned throughput will need to be redistributed among the existing partitions. For instance, if a table initially had 10 partitions, each with 2,000 RCUs and 500 WCUs, and you add 10 more, the throughput of each partition would be halved to accommodate the new partitions.

Let's now understand how bursting can help with throughput management, with unevenly distributed reads and writes in a DynamoDB table.

Bursting – short-term overprovisioning

In real-world scenarios, applications or users can disproportionately access certain keys, causing uneven distribution of requests across partitions. Bursting is a strategy to temporarily tap into any unused throughput from neighboring partitions to manage these short-term spikes in demand.

When allowing for bursting, it's essential to ensure workload isolation so that the extra throughput of one partition does not interfere with the regular operations of its neighboring partitions. This ensures that short-term gains in one partition do not compromise the overall system's performance. *Figure 5.8* shows the advantage of supporting bursting in DynamoDB, which allows it to serve more requests.

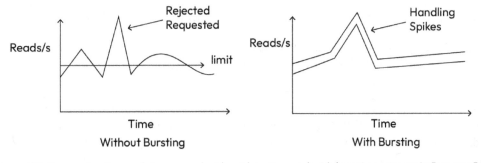

Figure 5.8: Comparing the reads/sec served without bursting and with bursting support in DynamoDB

A token bucket system

A token bucket system can be implemented at the node level to manage the bursting mechanism. Two buckets are maintained – the allocated throughput bucket and the available burst throughput bucket. If the regular bucket is empty (i.e., its provisioned throughput is exhausted), the system checks for available tokens in the burst bucket. If tokens are there, the partition is allowed to burst, temporarily exceeding its provisioned limits.

Bursting is beneficial for short-term uneven workload distribution, but if we need a more long-term approach to throughput management, DynamoDB design also supports adaptive capacity management, which we will cover next.

Adaptive capacity – long-term adjustments

While bursting handles short-term spikes in demand, adaptive capacity aims to reallocate throughput based on long-term usage patterns. This means that if certain partitions are consistently underutilized while others are overwhelmed, DynamoDB will gradually redistribute throughput to accommodate these patterns.

How it works

Under the hood, DynamoDB uses algorithms that study usage patterns over time, identifying hot partitions (partitions that are consistently accessed more frequently) and cold partitions (those less frequently accessed). A system then reallocates RCUs and WCUs accordingly. This ensures that your read and write operations are more evenly distributed, reducing the likelihood of throttling on busy partitions.

Adaptive capacity is effective but not instantaneous. It may take some time for a system to learn the access patterns and reallocate resources. Also, there is an upper limit to how much a single partition's throughput can be increased.

In the next section, we will learn about global admission control.

Global admission control – cross-partition management

Global admission control is a technique used to manage throughput across all partitions. While adaptive capacity focuses on individual partitions, global admission control takes a more holistic approach, managing resources at the table level.

One approach is to set a global limit on the number of operations per second, distributing this limit among partitions based on their load. This ensures that no single partition overwhelms a system, providing a more balanced throughput distribution.

Splitting for consumption – proactive partition management

If you anticipate a drastic change in the load pattern, you might decide to manually split or merge partitions to prepare for it. This is called "splitting for consumption."

For splitting, you could use key range partitioning or hash partitioning, depending on your data distribution and access patterns. The aim is to redistribute data such that each partition gets an equal share of the load, thereby maximizing throughput efficiency.

In conclusion, optimizing throughput allocation in a partitioned DynamoDB database involves multiple layers of strategies, each with its unique advantages and limitations. By understanding and judiciously applying these methods, you can significantly enhance the performance, reliability, and efficiency of your database operations.

In the next section, we will cover how DynamoDB is designed for high availability for reads and writes.

High availability in DynamoDB

High availability is a cornerstone of any large-scale database architecture, and DynamoDB is no exception. In this section, we will explore the various aspects that contribute to the high availability of reads and writes in DynamoDB. Let's first discuss the high availability of write requests in DynamoDB.

Write availability

DynamoDB's architecture adopts a partitioned model where tables are divided into partitions, and each partition is further replicated. These replicas of the same partition are termed a **replication group**. Leadership among these replicas is determined via a multi-Paxos-based leader election process. We covered the basics of leader election and multi-Paxos in *Chapter 3*. The leader replica manages all write requests by first logging them into a write-ahead log and then indexing them into memory. The leader replica later disseminates this write-ahead log and tree index to other replicas within the replication group.

The key to write availability is ensuring that a sufficient number of healthy replicas are available to form a write quorum. One robust strategy to maintain this availability is for the leader replica to promptly recruit another member into the replication group if it detects that a replica has become unresponsive or faulty.

For instance, let's consider four replication groups. Nodes from *group 3* may also be a part of *group 2*. If a node in replication group 4 becomes faulty but two-thirds of the nodes are still operational, it may appear that a quorum can still be formed. However, if the leader replica itself fails, achieving a quorum becomes impossible. This highlights the critical role of a healthy leader, as it's responsible for both processing writes and coordinating the election of a new leader if there are replica failures.

Now, let's discuss eventually consistent reads in DynamoDB.

Read availability

Read availability in DynamoDB is gauged by its ability to consistently return the most recent write upon a read request. DynamoDB's replication system offers eventual read consistency, with instant consistent reads provided solely by the leader replica. Therefore, it is crucial to ensure the leader replica's health for consistent read availability.

In DynamoDB, much hinges on the leader replica's reliability. If a leader fails, a new leader cannot take over until the previous leader's lease expires. To preemptively address this, the system should have rapid and accurate failure detection mechanisms. This is complicated by "gray failures," which are not straightforward to identify. A reliable way to counter gray failures is to establish communication protocols among replicas to confirm the leader replica's status before initiating a leader election.

By adopting these practices and philosophies, you can ensure that your DynamoDB database remains highly available, scalable, and efficient, thereby meeting the needs of your ever-evolving application landscape.

In conclusion, DynamoDB is a robust, serverless NoSQL database offered by AWS, designed for high performance, scalability, and availability. Its flexible, schema-less architecture accommodates a wide range of applications and workflows. The database efficiently manages throughput via mechanisms such as partitioning, bursting, and adaptive capacity. It also ensures high availability through replication groups and leader election processes for write and read operations. Overall, DynamoDB's multilayered strategies for throughput optimization and high availability make it a reliable, efficient choice for any large-scale, low-latency application.

Column-family databases

Column-family databases, a type of NoSQL database, are designed to handle vast amounts of data while providing high performance and scalability. They are particularly well-suited for applications that require the storage and retrieval of large volumes of data with high write and read throughput. Column-family databases are often used in distributed and horizontally scalable architectures. Apache Cassandra is one of the most prominent column-family databases.

Here are some key characteristics and concepts of column-family databases:

- **Data organization**: Data in a column-family database is organized into column families, which are essentially groups of related columns. Each column family contains a set of columns, and these columns can be dynamically added to the column family. Column families provide a flexible way to structure and store data.

- **Columns**: Columns within a column family are individual data elements. In column-family databases, columns do not need to be pre-defined, allowing you to add columns to the family as needed. This dynamic approach makes them suitable for handling evolving data requirements.

- **Rows**: Rows in a column-family database contain data specific to a particular entity or record. Data within a row is organized based on the column families associated with that row.

- **A wide-column store**: Column-family databases are sometimes referred to as "wide-column stores" because they can efficiently store a large number of columns for each row. This makes them suitable for applications that need to handle a wide variety of data attributes for each entity.

- **Scalability**: Column-family databases are designed for horizontal scalability. They can handle large volumes of data and high traffic by adding additional nodes to a cluster, and distributing data across multiple servers.

- **High write and read throughput**: These databases are optimized for high write and read throughput, making them suitable for real-time applications where data is constantly updated and retrieved.

- **Data distribution**: Data is distributed across nodes in the cluster, and replication is often employed to ensure fault tolerance and data availability:

  - **Querying**: While column-family databases excel in write-heavy and read-heavy workloads, they are not as suitable for complex querying compared to relational databases. Queries are typically optimized for key-based lookups.

  - **Use cases**: Column-family databases are commonly used in applications where scalability, high availability, and fault tolerance are essential, such as time-series data storage, event logging, monitoring systems, and distributed applications.

  - **Consistency models**: Column-family databases often provide tunable consistency levels, allowing you to balance between data consistency and system performance according to your application's requirements.

Apache Cassandra is a widely known open-source column-family database. It is particularly popular for its ability to handle large datasets distributed across multiple nodes, making it a valuable tool in applications that require high availability and seamless scalability. HBase is another column-family database that is highly consistent and sacrifices availability to maintain high consistency, as per the CAP theorem.

HBase

Apache HBase is an open-source, distributed, and scalable NoSQL database management system that is designed to handle large volumes of data with high read and write throughput. It is built on top of the **Hadoop Distributed File System** (**HDFS**) and was inspired by Google Bigtable. HBase is known for its ability to provide random and real-time access to massive amounts of structured data, making it suitable for applications with high data requirements, such as those found in big data and distributed computing ecosystems.

In short, in HBase, the following applies:

- A table is a collection of rows

- A row is a collection of column families

- A column family is a collection of columns

- A column is a collection of key-value pairs

The key features and characteristics of Apache HBase include the following:

- **Distributed and scalable**: HBase is designed to run on clusters of commodity hardware, and it offers horizontal scalability. It can handle the storage and processing of extremely large datasets by adding more nodes to a cluster.

- **A column-family data model**: Similar to Cassandra, HBase uses a column-family data model. Data is organized into column families, and each column family can contain a flexible number of columns.

- **A consistency model**: HBase offers strong consistency, making it suitable for applications that require strict data consistency. It uses a distributed **write-ahead log** (**WAL**) to ensure durability and consistency.

- **Data versioning**: HBase supports data versioning, allowing you to access and query previous versions of data, which can be valuable for historical analysis.

- **Scalability and load balancing**: HBase can automatically handle the distribution of data and load balancing across nodes, ensuring the efficient use of resources.

- **High write and read throughput**: HBase is designed for high write and read throughput, making it a popular choice for real-time data processing and analytics.

- **Hadoop integration**: HBase integrates well with the Hadoop ecosystem, and it can be used in conjunction with tools such as Hadoop MapReduce, Apache Spark, and Apache Hive for data processing and analysis.

- **Bloom filters and block caches**: HBase uses data structures such as bloom filters and BlockCaches to optimize data retrieval and improve query performance.

- **Compression**: Data compression is supported to reduce storage requirements and improve performance.

- **Use cases**: HBase is commonly used in applications that require random access to large datasets, such as time-series data, sensor data, log data, and applications related to internet services, including ad targeting and recommendation engines.

- **A community and ecosystem**: As an Apache project, HBase has an active open-source community and a rich ecosystem of tools, libraries, and connectors that support its development and usage.

HBase is particularly well-suited for applications that need to store and query large volumes of data in real-time, often with complex data models and high scalability requirements. Its integration with Hadoop and other big data technologies makes it a valuable choice for applications in the big data and analytics domains.

HBase details

Let's look at the design, architecture, and components of Hbase in detail. The concepts we will discuss here are inspired by the book *HBase – the Definitive Guide*.

HBase concepts and architecture

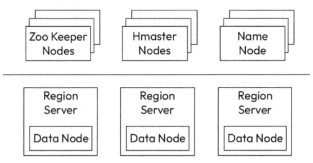

Figure 5.9: A diagram showing the different components of HBase architecture, namely the RegionServer co-located with the DataNode, the NameNode, the ZooKeeper, and the HBase Master nodes

HBase comprises three types of servers – RegionServer, HBase Master, and ZooKeeper:

- **RegionServer**: The RegionServer plays a vital role in handling data for both read and write operations. Clients directly interact with HBase RegionServers when accessing data, and the data it manages is stored in the Hadoop DataNode.

  To enhance data locality, RegionServers are co-located with HDFS DataNodes. This ensures that data is situated close to where it is required, optimizing efficiency. HBase tables undergo horizontal division into "regions," based on row key ranges.

  Each region spans the range from its start key to its end key, covering all rows within that boundary. These regions are then assigned to cluster nodes, known as "RegionServers," which efficiently serve data for both read and write operations. A RegionServer has the capacity to handle approximately 1,000 regions. Although the default size of a region is 1 GB, it remains configurable to suit specific requirements.

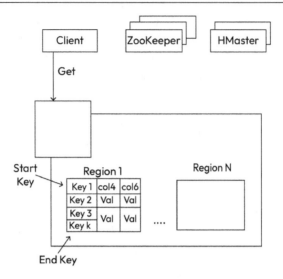

Figure 5.10: A diagram showing the internals of the RegionServer,
with the multiple regions it's responsible for

- **HBase Master**: The HBase Master process takes charge of tasks such as region assignment and DDL (Data Definition Language)operations (such as table creation and deletion). In a parallel fashion, the NameNode is responsible for maintaining metadata information related to all the physical data blocks that constitute the files. Specifically, the HBase Master oversees region assignment and DDL operations (CREATE, ALTER, TRUNCATE, and DROP).

HBase Master has several key responsibilities:

- The coordination of region servers

- Assignment of regions during startup, and the reassignment of regions for recovery or load balancing

- Monitoring all instances of a RegionServer within the cluster, attentively receiving notifications from a ZooKeeper

- Serving as an interface for crucial table operations, including the creation, deletion, and updating of tables

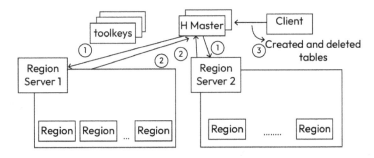

Figure 5.11: A diagram showing the role of HBase Master in the HBase architecture

ZooKeper: A ZooKeeper actively manages the real-time state of a cluster. In HBase, ZooKeeper functions as a distributed coordination service, ensuring the up-to-date status of servers in a cluster. It keeps track of the availability of servers, offering notifications if there are server failures. Utilizing a consensus approach, ZooKeeper guarantees a shared state among servers. It's important to note that a consensus is typically achieved with the involvement of three or five machines.

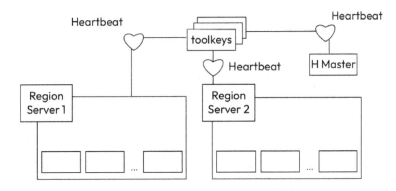

Figure 5.12: Shows the role of the ZooKeeper in HBase architecture

META table and .META. server

The META table in HBase, also known as the HBase Catalog table (as shown in *Figure 5.13*) stores information about the locations of regions within the cluster. The designated .META. server, known by the ZooKeeper, manages this special table. The META table functions as an HBase table, maintaining a comprehensive list of all regions in a system.

The structure of the .META. table is organized as follows:

Key: Consisting of the region's start key and its unique region ID

Values: Indicating the associated RegionServer

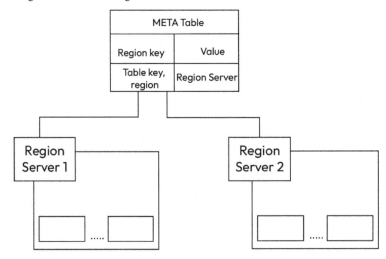

Figure 5.13: This diagram shows the META table and .META. server

RegionServer components

A Region Server, operating on an HDFS data node, comprises the following components (as shown in *Figure 5.14*):

- **WAL**: The WAL is a file on the distributed filesystem, used to store new data that is yet to be permanently persisted. Its primary purpose is to facilitate recovery if there is a failure.

- **BlockCache**: Functioning as the read cache, the BlockCache stores frequently read data in memory. When the cache reaches its capacity, the **Least Recently Used (LRU)** data is evicted.

- **MemStore**: Serving as the write cache, the MemStore stores new data awaiting a disk write. It undergoes sorting before being written to disk. There is one MemStore per column family per region, and updates are stored in memory as sorted KeyValues, mirroring their storage in an HFile.

- **HFiles**: These files store rows as sorted KeyValues on disk. Data is organized in HFiles, each containing sorted key/values. When the MemStore accumulates sufficient data, the entire set of sorted KeyValues is sequentially written to a new HFile in HDFS. This sequential write is exceptionally fast, as it eliminates the need to reposition the disk drive head.

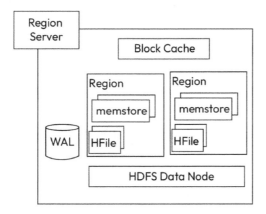

Figure 5.14: Shows the several components inside the RegionServer
– namely, BlockCache, WAL, MemsStore, and HFiles

First HBase access (read or write)

Let's look at the steps you need to execute when you access HBase for the first time in a read or write operation?"(as illustrated in *Figure 5.15*):

1. The client initiates communication with the ZooKeeper to retrieve details about the RegionServer, commonly referred to as the .META. server, hosting the META table.

2. Subsequently, the client queries the .META. server to obtain information about the specific region server associated with the desired row key.

3. The client then caches this information, including the location of the META table.

4. With this cached information, the client proceeds to retrieve the row from the pertinent RegionServer.

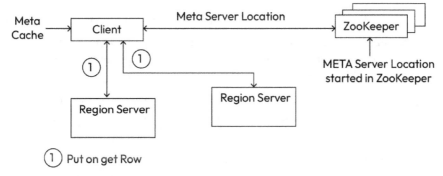

Figure 5.15: This diagram shows the first HBase read or write access

In subsequent read operations, the client relies on the cache to access the META location and row keys that were previously retrieved. As time progresses, querying the META table becomes unnecessary unless there is an occurrence of a miss, due to a region relocation. In such cases, the client will re-query the META table and update the cache accordingly.

HBase writes

When a client issues a `Put` request, the first step is to write data to the WAL. This is depicted in *Figure 5.16*:

1. Edits are appended to the end of the WAL file that is stored on disk. The WAL is used to recover not-yet-persisted data if a server crashes. The WAL is in an HDFS/filesystem outside the RegionServer.

2. Once the data is written to the WAL, it is placed in the MemStore. Then, the `put` request acknowledgment returns to the client.

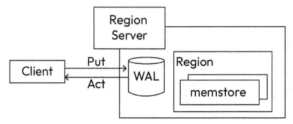

Figure 5.16: HBase writes are first written to WAL and then to
Memstore, and then they are sent back to the user

We will cover the HBase region flush in the next section.

HBase region flush

- When the MemStore accumulates enough data, the entire sorted set is written to a new HFile.

- HBase uses multiple HFiles per column family, which contain the actual cells, or KeyValue instances. These files are created over time as KeyValue edits sorted in the MemStores are flushed as files to disk.

- Hbase also saves the last written sequence number so that the system knows what has persisted so far. The highest sequence number is stored as a meta field in each HFile, reflecting where persistence has ended and where to continue from.

- On region startup, the sequence number is read, and the highest is used as the sequence number for new edits.

HBase reads

So, when you read a row, how does a system get the corresponding cells to return?

The KeyValue cells corresponding to one row can be in multiple places:

- Row cells already persisted are in HFiles
- Recently updated cells are in the MemStore
- Recently read cells are in the BlockCache

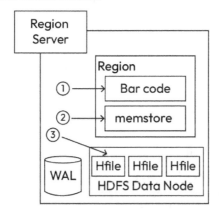

Figure 5.17: This diagram shows the flow of an HBase read – reading data from the BlockCache, MemStore, and the HFiles and then consolidating them before returning to the client

So, as shown in *Figure 5.17*, the read needs to check at all these places and do a merge by reading the key values from the BlockCache, MemStore, and HFiles in the following steps:

1. First, the scanner looks for the row cells in the BlockCache (the read cache). Recently read key values are cached here, and the least recently used are evicted when memory is needed.

2. Next, the scanner looks in the MemStore, the write cache in memory containing the most recent writes.

3. If the scanner does not find all of the row cells in the MemStore and BlockCache, then HBase will use the BlockCache indexes and bloom filters to load HFiles into memory, which may contain the target row cells.

Graph-based databases

Graph-based databases are a type of NoSQL database designed specifically to manage and store data with complex and interconnected relationships. They are well-suited for applications that require the modeling and querying of data in a way that reflects the relationships between various entities. These databases use graph structures to represent data and the connections between data points, making them highly efficient for traversing and querying complex relationships. Neo4j is one of the most well-known and widely used graph-based NoSQL databases.

Here are some key characteristics and concepts of graph-based NoSQL databases:

- **Graph structure**: Data in graph-based NoSQL databases is organized in the form of nodes and edges, creating a graph structure. Nodes represent entities (such as people, products, or locations), and edges represent the relationships between these entities.

- **Nodes**: Nodes are the fundamental units of data in a graph. Each node can contain properties (key-value pairs) that provide information about the entity it represents.

- **Edges**: Edges connect nodes and represent the relationships between them. Edges can also have properties to convey information about the nature of the relationship.

- **Labels and relationship types**: Nodes and edges can be labeled to group them into categories or types. For example, nodes representing people could be labeled as "person," and edges representing friendship could be labeled as "friends."

- **Traversal**: Graph-based databases are optimized to traverse relationships between nodes. This makes them highly efficient for queries that involve finding paths, connections, or patterns in data.

- **Cypher query language**: Graph-based databases often use the Cypher query language, specifically designed for querying and manipulating graph data. Cypher allows users to express complex graph queries in a human-readable and intuitive format.

- **Indexing**: Graph databases use indexing mechanisms to optimize query performance, allowing for fast lookups based on specific properties or relationship types.

- **Scalability**: Some graph databases offer horizontal scalability, enabling the distribution of data across multiple nodes for performance and fault tolerance.

- **Use cases**: Graph-based databases are commonly used in applications that require modeling and analyzing complex relationships, such as social networks, recommendation engines, fraud detection, network and infrastructure management, and knowledge graphs.

- **Pattern matching**: These databases excel at pattern matching and can find patterns and connections within the graph, enabling applications such as social network friend suggestions and personalized recommendations.

Graph-based NoSQL databases are valuable tools for scenarios where understanding and querying the relationships between data points are critical. They are particularly well-suited for applications that involve navigating and analyzing intricate and evolving networks of data. For example, consider a use case where we need to find and track friendship and follower recommendations in a social network application.

A graph database can quickly traverse the graph to suggest friends of friends (second-degree connections) or people with similar interests based on common interactions and connections.

Neo4j is an example of a graph-based NoSQL database. Let's take a deeper dive into Neo4j in the next section.

The Neo4j graph database

Neo4j is a popular and widely used graph **database management system** (**DBMS**), known for its ability to efficiently store, manage, and query data with complex relationships. It is specifically designed for applications that require modeling, storing, and traversing intricate and interconnected data structures. Neo4j is often used in scenarios where understanding and querying relationships between data points are fundamental, making it well-suited for a wide range of applications, including social networks, recommendation engines, fraud detection, network and infrastructure management, and knowledge graphs.

The key features and characteristics of Neo4j include the following:

- **A graph data model**: Neo4j employs a graph data model, representing data as nodes and the relationships between nodes. This model provides an intuitive way to express and manage complex relationships.

- **Nodes**: Nodes in Neo4j represent entities or data points. Each node can have properties (key-value pairs) that describe the attributes of the entity it represents.

- **Relationships**: Relationships between nodes define connections and associations between data points. Relationships can also have properties, providing additional information about the nature of the relationship.

- **Labels and relationship types**: Nodes and relationships can be labeled to group them into categories or types. For example, nodes representing people could be labeled as "person," and relationships representing friendships could be labeled as "friends."

- **Cypher query language**: Neo4j uses the Cypher query language, specifically designed for querying and manipulating graph data. Cypher is known for its readability and expressiveness when dealing with complex graph structures.

- **Indexing and query optimization**: Neo4j uses indexing mechanisms to optimize query performance, allowing for efficient lookups based on specific properties or relationship types.

- **Scalability**: Neo4j supports horizontal scalability, enabling the distribution of data across multiple nodes to enhance performance and fault tolerance.

- **ACID compliance**: Neo4j ensures data integrity by adhering to ACID properties, which are crucial for maintaining data consistency.

- **Data versioning**: Neo4j can store and query historical data, making it valuable for scenarios requiring a temporal view of relationships.

- **Use cases**: Neo4j is commonly used in various applications, including social networks, recommendation systems, fraud detection, real-time analytics, network and infrastructure management, and knowledge graphs.

- **Community and ecosystem**: Neo4j has an active open source community and offers a rich ecosystem of tools, libraries, and connectors that support its development and usage.

Neo4j's strength lies in its ability to efficiently navigate and query complex relationships in the data, making it a powerful choice for applications where understanding and exploiting data connections are essential. Its ease of use, scalability, and support for graph-based analytics have made it a leading choice in the field of graph databases.

Neo4j details

Let's understand the data modeling and inner workings of the Neo4j database by taking a simple example. Consider a simple social network graph, where there are many users and they follow each other. In this example (as shown in *Figure 5.18*), there are three people, represented by three nodes (nodes 1, 2, and 3). **Node 1** follows 2 and 3. **Node 2** follows 3.

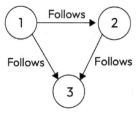

Figure 5.18: The relationship between three nodes

Relational modeling versus graph modeling

We have two options to store the data if we adopt a relational modeling point of view:

- Store the source and destination of an edge (relationship) as a row, and do that for all the edges, as shown in **(a)** in *Figure 5.19*. When you need to find the outgoing edges (whom a particular user follows), you can use an index to "seek" the start of that user's outbound relationships.

- Store an edge twice and add another column to indicate the direction, as shown in **(b)** in *Figure 5.19*.

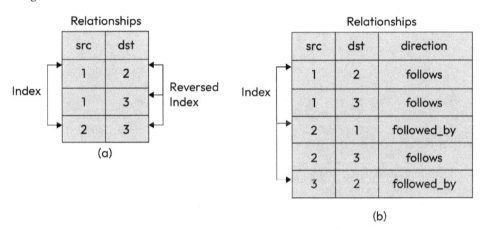

Figure 5.19: A diagram showing the options to store graph data using a relational model

Graph modeling

Now, if we want to store data using graph modeling, we would represent the data in doubly linked list graphs and two tables, called `nodes` and `relationships`:

- *Figure 5.20* **(a)** represents nodes and their relations as doubly linked lists. We have three nodes, hence three linked lists.

- *Figure 5.20* **(b)** is the actual node storage. Each record stores the **first relationship ID (first_rid)** of a node. For example, the first relationship for both nodes 1 and 2 is A.

- *Figure 5.20* **(c)** is the actual relationship storage. Each record stores the **source (src)** and **destination (dst)** of a relationship. In addition, it also stores the previous and next relationship IDs for both the source and destination nodes (`src_prev_rid`, `src_next_rid`, `dst_prev_rid`, and `dst_next_rid`).

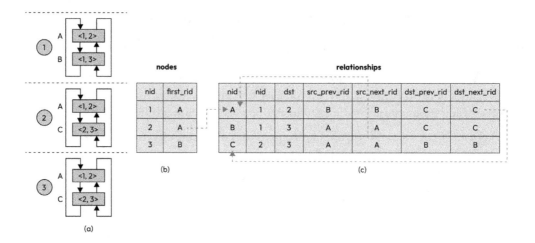

Figure 5.20: A diagram showing a representation of graph data using graph modeling

Adding a new node to an existing graph

If you want to add **node 4** to the graph (as shown in *Figure 5.21* **(a)**), where **node 2** follows **node 4**, you would make changes to the doubly linked lists and the two tables, as follows.

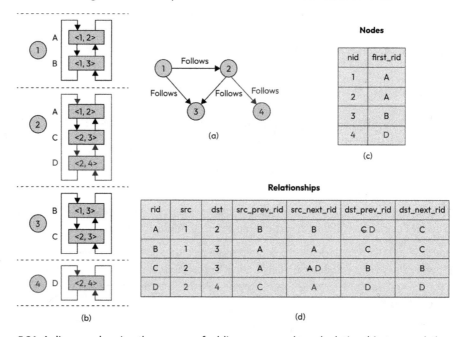

Nodes

nid	first_rid
1	A
2	A
3	B
4	D

(c)

Relationships

rid	src	dst	src_prev_rid	src_next_rid	dst_prev_rid	dst_next_rid
A	1	2	B	B	C D	C
B	1	3	A	A	C	C
C	2	3	A	A D	B	B
D	2	4	C	A	D	D

(b)   (d)

Figure 5.21: A diagram showing the process of adding a new node and relationship to an existing graph

The new node's `nid` is 4, and we added a new relationship D, where we show that `src` is 2 and `dst` is 4, i.e., node 2 follows node 4.

In the nodes table, you would add an entry with `nid = 4` and `first_rid = D`.

In the relationship table, you would add an entry for the new relationship, D, with the appropriate entries for the columns (`rid`, `src`, `dst`, `src_prev_rid`, `src_next_rid`, `dst_prev_rid`, and `dst_next_rid`) and make changes to the existing entries, as shown in the figure.

Nodes and the relationship store

Node store

All the nodes in the database are stored in the node store file.

Each node record accounts for a fixed 15 bytes.

The layout is as follows:

- `1st byte — isInUse`
- `Next 4 bytes — ID of the node`
- `Next byte — First relationship ID`
- `Next byte — First property ID`
- `Next 5 bytes — Label store`
- `Remaining byte — for future use`

Relationship store

Each relationship record is a fixed record of 34 bytes.

The relationship record's layout is as follows:

- `Start node ID`
- `End node ID`
- `Pointer to the relationship type`
- `Pointer to the next and previous relationship record for each of the start node and end node`

Summary

In this chapter, we traversed the landscape of databases, exploring the fundamental concepts and diverse array of database types that play a pivotal role in modern data management.

We began by unraveling the essence of databases themselves, recognizing their crucial role in data organization, retrieval, and integrity. Databases serve as the foundation to manage data in various applications and systems, facilitating informed decision-making and efficient data processing.

Our exploration then delved into the world of database types, where we encountered two primary categories – relational databases and NoSQL databases. Relational databases, with their tabular structure and strong data integrity, are well-suited for applications demanding structured data and complex relationships. Notable examples include MySQL, PostgreSQL, and Oracle.

At the other side of the spectrum, we embraced the diversity of NoSQL databases, which offer flexibility and scalability. Within this category, we uncovered key-value stores, which provide efficient data retrieval and storage, and specifically explored Amazon DynamoDB as an example of this type. We navigated the landscape of column-family databases, where Apache Cassandra and HBase emerged as scalable solutions to handle extensive data volumes and write-heavy workloads. We took a deep dive into the HBase design, architecture, and inner workings.

Our journey continued to the realm of graph-based databases. These databases excel in representing and navigating complex relationships, making them invaluable for applications with intricate network structures, such as social networks, recommendation engines, and knowledge graphs. We specifically looked into Neo4j as an example of the graph database and explored it in a bit more detail.

In conclusion, our chapter traversed the spectrum of databases, from the structured world of relational databases to the versatile landscape of NoSQL, encompassing key-value stores, column-family databases, and graph-based databases. This understanding empowers us to make informed choices when selecting the right database type for the unique demands of each application and data scenario.

In the next chapter, we will look at caches and their types and purposes in software system design.

References

HBase – The Definitive Guide: `https://www.oreilly.com/library/view/hbase-the-definitive/9781449314682/`

6

Distributed Cache

In the rapidly evolving landscape of modern computing, the demand for scalable, high-performance systems has become paramount. As applications grow in complexity and user bases expand, traditional approaches to data retrieval and storage may encounter bottlenecks that hinder overall performance. One of the most effective strategies to enhance performance is the implementation of caching. Caching involves temporarily storing copies of data in locations closer to the user or application, thereby reducing access time and improving efficiency. At its core, caching is a technique used to store frequently accessed data in a high-speed data storage layer. This storage layer, or cache, can be located in various places, such as in memory (RAM), on a disk, or even in a network. By keeping a subset of data in these faster access locations, systems can significantly reduce the latency experienced when fetching data, leading to improved application performance and user experience.

As applications scale and distribute across multiple servers or data centers, a single cache may no longer suffice. This is where distributed caching comes into play. Distributed caching involves using a network of cache nodes that work together to provide a cohesive caching layer across multiple servers or locations.

This chapter delves into the principles of caching and the more advanced concept of distributed caching, exploring their importance, mechanisms, and applications in modern computing.

We will be covering the following concepts in detail in this chapter:

- What is caching?
- What is distributed caching?
- Designing a distributed cache
- Popular distributed cache solutions

Let's start by looking at caching.

What is caching?

Before we start digging deeper into distributed caching, let's understand caching. Caching is a technique used in computing to store and manage copies of data or resources in a location that allows for faster access. The primary purpose of caching is to reduce the time and resources required to retrieve data by keeping a copy of frequently accessed or expensive-to-compute information in a readily accessible location. This location is typically faster to access than the original source of the data.

Here are the key concepts associated with caching:

- **Cache**: A cache is a temporary storage area that holds copies of frequently accessed data or resources. This can be in the form of a hardware cache (for example, a CPU cache) or a software-based cache (for example, an in-memory cache).

- **Cached data**: This refers to the copies of data that are stored in the cache. The data is usually obtained from a slower, more permanent storage location (such as a database or disk) and is kept in the cache for faster retrieval.

- **Cache hit**: A cache hit occurs when the requested data is found in the cache. This results in a faster retrieval process since the data is readily available without the need to go to the original data source.

- **Cache miss**: A cache miss happens when the requested data is not found in the cache. In this case, the system needs to fetch the data from the original source and store a copy in the cache for future access.

- **Eviction**: In situations where the cache has limited space, eviction may occur when the cache is full and needs to make room for new data. The system may remove less frequently accessed, less recently used, or older data to accommodate new entries.

Caching plays a crucial role in optimizing system performance, reducing latency, and improving the overall user experience in a wide range of computing environments.

What is distributed caching?

Distributed caching is a technique employed to optimize data access by strategically storing frequently accessed information in memory across multiple interconnected servers or nodes. Rather than repeatedly fetching the same data from the primary data source, a distributed cache ensures that a copy of this data is readily available in the cache, significantly reducing latency and enhancing system responsiveness.

The primary objective of distributed caching is to mitigate the performance challenges associated with accessing data from slower, disk-based storage systems. By maintaining a cache in the main memory of multiple nodes, the system can quickly retrieve frequently accessed data without incurring the delays associated with disk I/O operations. This caching strategy proves particularly effective in scenarios where rapid access to data is critical, such as in web applications, databases, and other data-intensive environments.

How is it different from regular caching?

Distributed caching and regular (non-distributed, local, single node) caching both involve the storage and retrieval of frequently accessed data to improve system performance. However, they differ in terms of their scope and architecture, as well as the scale at which they operate. Here are the key distinctions between distributed caching and regular caching:

	Regular caching	Distributed caching
Scope	Caching typically refers to the practice of storing and retrieving frequently accessed data within a single local cache. This cache could be a part of the application or system and may exist on a single machine or server.	Distributed caching extends the concept of caching to a network of interconnected nodes or servers. In a distributed caching system, the cache is spread across multiple machines, enabling the sharing of cached data among these nodes.
Architecture	In a traditional caching setup, data is stored in a local cache, which is often located in memory. The cache is directly accessible by the application running on a single machine.	Distributed caching involves a network of cache nodes, where each node may have its local cache. These nodes communicate and collaborate to share cached data. The architecture is designed to scale horizontally by adding more nodes to the distributed environment.
Scale	Caching is suitable for smaller-scale scenarios, such as a single server or a standalone application. It is effective when the performance improvement gained from caching on a local machine is sufficient.	Distributed caching is designed to address the challenges of larger-scale systems and applications. It is particularly beneficial in scenarios where data needs to be shared and accessed across multiple nodes to achieve improved performance and scalability.
Use cases	Common use cases for caching include improving the performance of local applications, reducing database access times, and speeding up the retrieval of frequently accessed resources within a single server or application.	Distributed caching is applied in scenarios where the scale and distribution of data access require a more coordinated approach. It is commonly used in large-scale web applications, distributed databases, and microservices architectures.
Consistency and coordination	In the regular caching (single-node) scenario, maintaining cache consistency is relatively straightforward. However, cache invalidation and ensuring data coherence in a distributed environment can be more complex.	Distributed caching systems implement mechanisms for maintaining consistency across nodes, ensuring that all nodes have access to the most up-to-date and synchronized data. Coordination protocols and distributed cache management strategies are employed to handle potential challenges.

Table 6.1: Regular caching versus distributed caching

While both regular and distributed caching aim to improve system performance by storing frequently accessed data, distributed caching extends the concept to a network of interconnected nodes, addressing the challenges of larger-scale and distributed computing environments. The choice between regular caching and distributed caching depends on the specific requirements and scale of the application or system being designed.

Use cases

Let's look at some of the use cases of distributed caching

- **Web applications**: Distributed caching is extensively used in web applications to store frequently accessed data such as user sessions, page fragments, and database query results

- **Database query results**: Caching frequently executed database queries or query results helps reduce the need for repeated database access

- **Content Delivery Networks (CDNs)**: CDNs leverage distributed caching to store and serve static content (images, videos, stylesheets) at strategically located edge servers

- **Session management**: Storing session data in a distributed cache allows for efficient and scalable session management in web applications

- **API response caching**: Caching API responses helps reduce the load on backend servers and speeds up the delivery of frequently requested data

- **Real-time analytics**: Caching aggregated or frequently queried analytics data enables faster retrieval for real-time reporting and dashboard generation

- **Message queues**: Caching message queues can improve the efficiency of message processing systems by storing intermediate results

Benefits of using a distributed cache

Here are some of the benefits of using the distributed cache:

- **Performance improvement**: Distributed caching significantly reduces data access times by keeping frequently accessed data in memory. This results in faster response times and improved system performance.

- **Scalability**: As system demands grow, distributed caching allows for easy scaling by adding more cache nodes. This ensures that the cache can accommodate increased loads and maintain optimal performance.

- **Reduced load on backend systems**: By serving frequently accessed data from the cache, distributed caching minimizes the load on backend storage systems (such as databases and file systems), leading to more efficient resource utilization. The number of requests going to the backend is reduced. This can immensely help mitigate the risks of origin servers going down by an overwhelming margin in the event of a DDoS attack.

- **Fault tolerance**: Distributed caching systems often provide redundancy and fault tolerance. In the event of a node failure, other nodes can continue serving cached data, ensuring system reliability and continuity.

- **Consistent access times**: Caching ensures consistent and predictable access times for frequently requested data, regardless of the size or complexity of the overall data set.

- **Cost savings**: Improved performance and reduced load on backend systems can lead to cost savings in terms of infrastructure resources, as the need for additional servers or resources may be minimized.

- **Enhanced user experience**: Faster response times and improved system performance contribute to a more responsive and seamless user experience, which is crucial in applications and services where user satisfaction is paramount.

In summary, distributed caching proves invaluable in addressing performance bottlenecks, enhancing scalability and security, and optimizing resource utilization in a variety of computing scenarios. The specific benefits and use cases may vary based on the requirements of the application or system that is being designed.

Challenges of using distributed caching

While distributed caching provides numerous benefits, it also comes with certain potential drawbacks and challenges. It's important to be aware of these considerations when implementing distributed caching solutions.

Here are some potential drawbacks:

- **Consistency challenges**: Maintaining data consistency across multiple cache nodes can be challenging. Ensuring that all nodes have the most up-to-date information requires coordination mechanisms, and achieving perfect consistency may introduce latency or trade-offs.

- **Cache invalidation complexity**: Cache invalidation, the process of removing or updating cached data when the underlying data changes, can be complex in a distributed environment. Ensuring that all nodes are aware of changes and update their caches accordingly can introduce additional overhead.

- **Increased complexity and configuration**: Setting up and configuring a distributed caching system can be more complex than using a single-node caching solution. The need for coordination, partitioning strategies, and proper configuration settings adds complexity to the deployment and maintenance of the system.

- **Network overhead:** The communication between cache nodes introduces network overhead. In situations where nodes need to coordinate and share updates, network latency and bandwidth can become limiting factors, especially in geographically distributed systems.

- **Potential for cache staleness:** Due to the distributed nature of caching, there's a risk of cache staleness, where a node might serve outdated data if it hasn't received updates from other nodes. This issue can occur during cache expiration periods or when the cache is not properly synchronized.

- **Data partitioning challenges:** Distributing data across multiple cache nodes requires effective partitioning strategies. Poor partitioning decisions can lead to uneven distribution of data, resulting in some nodes being overloaded while others are underutilized.

- **High memory usage:** In scenarios where the cache needs to store large amounts of data, distributed caching systems may consume a significant amount of memory across multiple nodes. This can impact the overall system's resource usage and scalability.

- **Cost:** Implementing and managing a distributed caching solution may incur additional costs, both in terms of infrastructure (hardware or cloud resources) and the complexity of maintenance. The benefits should be carefully weighed against the associated costs.

- **Data access patterns:** Certain access patterns, such as random or infrequent access, might not benefit as much from distributed caching. In such cases, the overhead of maintaining a distributed cache may outweigh the performance gains.

- **Limited utility for write-intensive workloads:** Distributed caching is typically more beneficial for read-intensive workloads. Write-intensive workloads, where data is frequently updated, may face challenges in maintaining consistency and coordination across cache nodes.

It's essential to carefully evaluate the specific requirements of your application and consider these drawbacks when deciding whether to implement distributed caching. Additionally, choosing the right distributed caching solution, configuring it properly, and monitoring its performance can help mitigate some of these challenges.

In the subsequent sections, we will delve deeper into the intricacies of designing and implementing distributed caching solutions, exploring key considerations, architectures, and best practices that contribute to the successful integration of distributed caching into diverse computing environments.

Designing a distributed cache

We will now design a distributed cache. Let's start by noting down the requirements and then creating a high-level diagram with all the components. We will then go into a detailed design hashing out the inner workings of the components.

Requirements

Thinking about the requirements for the distributed cache, we can categorize them into two areas: functional and non-functional requirements.

The following are the functional requirements:

- `put`(key, value): We should be able to add a key and value pair to the cache
- `get`(key): Given the key, we should be able to fetch the corresponding value

The following are the non-functional requirements:

- **Highly performant**: The system should deliver fast and efficient access to cached data, providing low-latency responses and high throughput. Performance is a critical aspect of distributed caching, especially in scenarios where quick access to frequently used data is essential for improving overall system responsiveness.
- **Highly scalable**: The system should be capable of efficiently handling increased workloads and growing demands by adding more resources or nodes. This is important so that the system can handle a higher volume of data, requests, or users without a significant degradation in performance.
- **Highly available**: We should minimize downtime and ensure that the service remains accessible and operational even in the presence of failures or disruptions. This is a critical attribute for systems that require continuous accessibility to support mission-critical applications and services.

Design journey

Now that we have listed the requirements, let's build the solution iteratively by enhancing the design in each iteration. First, we will think about what is an appropriate data structure for the cache. Then we will look into the most suitable system arrangements and deployment strategies, and then we will evaluate whether the design satisfies the functional and non-functional requirements.

Data structure for a cache

What are our options for the underlying data structure that we will use to store and retrieve the cache entries? A very simple solution would be to use a HashMap (or a HashTable) that can store a key and corresponding value. HashMap data structure has constant time `put` and `get` operations and therefore is a good choice.

What if the HashMap size becomes more than the capacity of the server memory? This can be mitigated by adopting an eviction policy. Let's take a look at the various options we have.

Cache eviction policies

Cache eviction refers to the process of deciding which entries to remove from the cache when it reaches its capacity or when certain conditions are met. Let's explore some of the cache eviction policies.

- Insertion based, not accounting for access time:

  - **First in, First out (FIFO)**: Consider a restaurant reservation system that holds a limited number of reservations in a cache. When the cache is full and a new reservation is made, the oldest reservation (the one made first) is evicted to accommodate the new one. This ensures that reservations are managed in the order in which they were received, adhering to the FIFO principle.

  - **Last in, First out (LIFO)**: Social media story suggestions are an example of the LIFO principle. The news suggestion that was just shown to the user can be evicted when the cache is full to give a chance to stories that were not suggested recently.

- Access based:

  - **MRU**: The idea here is to evict the most recently used entry. Document editing software is an example. Imagine a piece of document editing software that keeps a cache of recently opened documents for quick access. If the cache size is limited and a new document is opened, the most recently used document (the one that was just accessed or edited) might be evicted to make room for the new one. This approach is useful in scenarios where the most recent items are considered the least likely to be needed again immediately, following the MRU eviction policy.

  - **Least Recently Used (LRU)**: This policy is very common. It evicts the least recently used entry and keeps the most recently used entries. A web browser cache is an example. Web browsers often use LRU cache policies for storing web pages. When a user visits a website, the page is cached. If the browser cache is full and a new page needs to be cached, the page that has not been accessed for the longest time is evicted. This ensures that frequently accessed pages are kept in the cache, improving browsing speed and performance.

  - **LFU**: This policy involves evicting the least frequently used entry. In a music streaming service, for example, a cache might be used to store the most frequently played songs for quicker access. If the cache becomes full, the song that has been played the least number of times is evicted to make room for a new song. This ensures that the cache holds the songs that users listen to most often, improving the efficiency and user experience of the service.

Eviction triggers can be size-based or days-based. If the cache has reached its capacity 'C', or if the entries are more than 'N' days old, the entries will be evicted based on one of the preceding logic policies.

Designing an LRU cache

As we see, there are many cache eviction policies out there, but a simple and popular eviction policy is to evict the LRU cache entry. Let's note down the high-level requirements for an LRU cache:

- There is a limited number of entries in the cache (let's call it N)
- We need to be able to add an entry to the cache in O(1)
- We need to be able to remove an entry to the cache in O(1)

Let's consider a doubly linked list and a HashMap combination data structure as shown in *Figure 6.1* and see whether this works well:

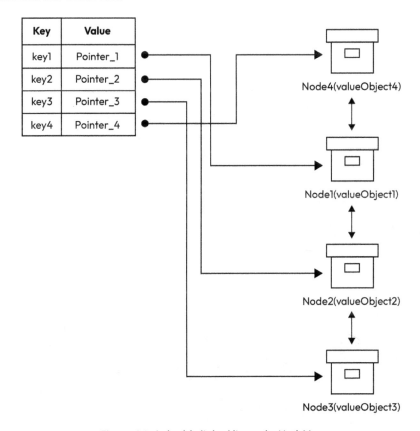

Figure 6.1: A doubly linked list and a HashMap

There are two flows here – a **read flow** and a **write flow**. For simplicity, let's assume that writes would go to the database directly. Only in the read time, if there is a cache miss, will the cache be populated.

Let us consider the different cases based on the key being present in the cache and not present in the cache.

So, the primary flow is a read flow – given a key, we need to find its corresponding `valueObject` (we will be referring to it as `valueObject` to differentiate between this "value" and the HashMap's value field). The `valueObject` is the actual data we are trying to cache here. This could be the profile of a user, for example, which would look like this:

```
valueObject: {
    "name": "John Doe",
    "city": "San Jose",
    "state": "California"
}
```

So, the first step is to check the `HashMap` to see whether the key is present. Here are a few possible scenarios:

- **Case 1**: The key is not present in the `HashMap`; it's a cache miss. So, it's a new entry. The number of entries in the cache is less than N. In this instance, follow these steps:

    - Fetch the `valueObject` for the key from the actual database

    - Create a node with the `valueObject` and add it in front of this linked list.

    - Add the corresponding entry in the HashMap with the key and value as the pointer to the node in the linked list.

- **Case 2**: The key is not present in the `HashMap`; it's a cache miss. So, it's a new entry, and the number of entries in the cache is equal to N. In this instance, you can follow these steps:

    - Since the cache is full, we need to create space.

    - Go to the end of the `LinekdList` and fetch the entry at the end, which is the least recently used entry.

    - Remove this entry from the `LinkedList` as well as from the HashMap

    Now, the number of entries is less than N. This case is exactly like in the first case, so follow those steps.

- Fetch the `valueObject` for the key from the actual database.

- Create a node with the `valueObject` and add it in front of this linked list.

- Add the corresponding entry in the HashMap with the key and value as the pointer to the node in the linked list.

- **Case 3**: The key is present in the `HashMap`, so It's an existing entry:

  - Locate the node in the `LinkedList` with this key by doing a lookup in the `hashMap` and following the pointer stored in the value field of the `HashMap`.

  - We need to move this node to the front of the `LinkedList`.

All of these operations can be done in O(1) time.

Putting the system together

Now that we have designed the data structure, let's go into the various arrangements for deploying the cache.

Solution 1 – a co-located cache solution

As shown in *Figure 6.2*, we can run the cache as a process and co-locate it in the same machine as the app server. To scale up, we would have as many cache instances as the app servers, since they are co-located. This approach is scalable too. Some details on the solution are as follows:

- **Advantage**: It will be quicker to do a cache lookup since it will be an interprocess call in the same machine.

- **Challenge**: What happens if the machine goes down? The new request would be handled by a new app server, but the cache would be empty. We may need to think about a cache solution where the cache is not co-located with the AppServer machine. There are trade-offs to be made here, but let's think about a standalone cache solution in the next section.

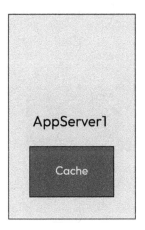 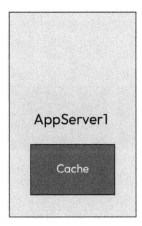 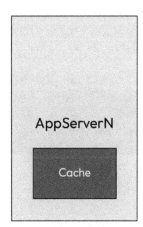

Figure 6.2: A co-located cache solution

Solution 2 – a standalone cache solution

As shown in *Figure 6.3*, we can deploy the cache on an independently scalable cluster of hosts. The details for this solution are as follows:

- **Advantage**: It can support the scale to any frequency and concurrency of requests. This can also scale independently of the number of app servers.

- **Challenge**: We may need to share the cache entries based on the cache key and put some sort of load balancer or lookup map to find which host contains the appropriate key. We can use a simple modulo approach to shard the keys, but a better strategy would be using a consistent hashing approach we discussed earlier in this book (in *Chapter 3*).

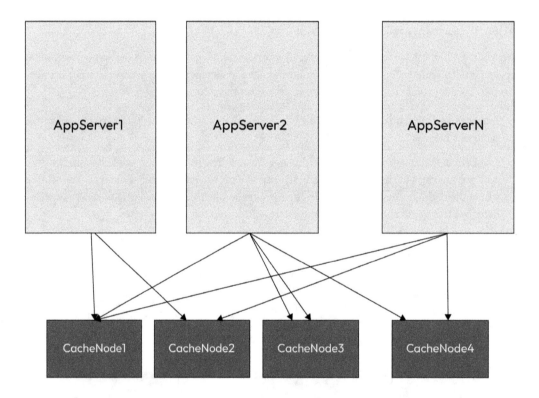

Figure 6.3: A standalone cache solution

Next, let us explore how can we choose between the two cache solutions.

How to choose between the two cache arrangements

Choosing between a co-located cache with the app server and a standalone cache involves considering various factors related to your application's requirements, architecture, and priorities. *Table 6.2* shows some key considerations to help you make an informed decision:

	Co-located cache	Standalone cache
Performance requirements	If low-latency access to cached data is crucial and the application demands fast response times, a co-located cache may be preferred due to direct access without network overhead.	If scalability and high throughput are more critical than minimal latency, a standalone cache that can be independently scaled might be a better choice.
Scalability	It is suitable for scenarios where the application and cache can scale together, and the scalability demands are not extremely high.	It offers better scalability as the cache can be scaled independently, allowing for more granular control over resource allocation.
Resource sharing	Shared resources with the app server can be efficient in terms of memory and CPU utilization.	It provides isolation, avoiding resource contention between the app server and cache, which can be beneficial for large-scale systems.
Flexibility	It comes with a simpler configuration and integration process but may have limitations in terms of technology choices and configurations.	It comes with more flexibility in choosing caching solutions based on specific requirements and the ability to select technologies independently.
Dependency and isolation	It is tightly coupled with the app server; changes or issues in one may impact the other.	It offers independence and isolation, reducing the risk of system-wide disruptions due to changes in one component.
Operational complexity	It offers simpler deployment and management but may lack the flexibility to address specific caching needs.	It may involve more complex configuration and management but allows for tailored solutions and updates without affecting the app server.
Infrastructure and cost	It may be more cost-effective in terms of infrastructure, as it shares resources with the app server.	It requires additional infrastructure, potentially leading to higher operational costs.
Network latency tolerance	It is well-suited for applications where minimizing network latency is a top priority.	It is acceptable if the application can tolerate slightly higher network latency for cache access.

Table 6.2: Key considerations to choose between a co-located or standalone cache

Ultimately, the choice between co-located and standalone caching depends on the specific needs of your application, and a careful evaluation of the trade-offs in terms of performance, scalability, flexibility, and operational considerations is essential. Additionally, considering future growth and changes in requirements can help ensure that the chosen caching architecture aligns with the long-term goals of your system.

Evaluate the design against the requirements

This seems like a good point to evaluate if we have met our functional and non-functional requirements. Functional requirements seem to be satisfied with the `put` and `get` functions working fine.

Let's look at the non-functional requirements.

- **Highly performant**: The operations are all O(1), so it's very efficient.

- **Highly scalable**: This seems to be fine since we can scale it to any number of requests and concurrency by scaling the number of hosts where this cache is deployed and sharding them properly.

- **Highly available**: This criterion seems to still not be fully satisfied. If one host goes down, then we lose the entire shard.

Let's dig in a bit more and think about a strategy to address the high availability requirement. We can achieve this via data replication – keeping multiple copies with each copy deployed into different hosts. So, if one host goes down, there will still be additional copies in different hosts that can serve the request. Typically, we would keep two additional copies, so there are three copies on three different hosts. One of them is a primary copy and the other two are secondary copies. The secondary copies can also be used to serve higher read traffic. Writes would go to the primary host first and then be replicated to secondary hosts.

This data replication arrangement introduces another challenge – consistency. There is a trade-off here as to whether to choose availability or consistency. We have discussed this trade-off in earlier chapters. So, we will not go into a lot of detail.

Popular distributed cache solutions

There are two very popular distributed cache solutions out there in the market – Redis and Memcached. Redis and Memcached are two widely used distributed cache solutions, prized for their speed and simplicity in efficiently storing and retrieving data. Understanding the strengths of Redis and Memcached is crucial for developers looking to optimize data access in distributed environments. Let's explore each one of them a bit more.

Redis

Redis is a versatile in-memory data store that supports a range of data structures such as strings, lists, sets, and hashes. Offering persistence options for durability, it goes beyond basic storage capabilities with advanced features including pub/sub messaging, transactions, and Lua scripting. This flexibility allows Redis to serve multiple purposes, functioning as a cache, message broker, or even a full-fledged database. Let's explore its common use cases, scalability, and community support.

Use cases

- Caching in web applications
- Real-time analytics
- Session storage
- Leaderboards and counting systems

Scalability

- Redis is horizontally scalable through sharding, allowing you to distribute data across multiple nodes.

Community and support

- Active open-source community with widespread adoption
- Robust documentation and community support

Memcached

Memcached stands out as a straightforward key-value store, functioning seamlessly as an in-memory caching solution. Known for its lightweight and user-friendly design, it efficiently supports simple data types while leveraging a distributed architecture to enhance its scalability. Let's see what use cases Memcached is good for, its scalability and the community support it has.

Use cases

- Caching in web applications.
- Session storage.
- Database result caching.
- Distributed systems where a simple key-value store is sufficient.

Scalability

- Memcached is designed to scale horizontally by adding more nodes to the cache cluster.
- Data is distributed across nodes using consistent hashing.

Community and support

- Well-established and widely adopted, with a mature code base
- Simple and easy to integrate, suitable for various programming languages

How to choose between Redis and Memcached

Here are some considerations for choosing between the two options:

- **Use case specificity**: The choice between Redis and Memcached often depends on specific use case requirements. Redis, with its versatile data structures, is suitable for a broader range of scenarios, while Memcached excels in simplicity and speed for basic key-value caching.
- **Persistence**: Redis provides options for persistence, making it more suitable for use cases requiring data durability. Memcached, being an in-memory cache, does not inherently provide persistent storage.
- **Data structure support**: Redis supports a wider range of data structures and features, making it more versatile in certain scenarios where complex data manipulation is required.
- **Ease of use**: Memcached is known for its simplicity and straightforward design, making it easy to integrate and operate. Redis, despite being more feature-rich, might have a steeper learning curve for some users.

Ultimately, the choice between Redis and Memcached depends on the specific requirements of your application, the complexity of data manipulation needed, and considerations such as ease of use and community support.

Summary

In this chapter, we started by defining caching as a computing technique to store and manage copies of data for faster access. The primary goal is to reduce the time and resources needed to retrieve frequently accessed or computationally expensive information. We then covered some other key concepts including cached data, cache hit, cache miss, and eviction policies

Then we delved into distributed caching. Distributed caching optimizes data access by strategically storing frequently accessed information across multiple interconnected servers or nodes. We learned that it aims to mitigate performance challenges related to slower, disk-based storage systems. This is effective in scenarios where rapid access to data is crucial, such as in web applications and databases. We explored the differences between caching and distributed caching in terms of the following dimensions: scope, architecture, scale, and use cases.

We talked about the benefits of distributed caching such as performance improvement, scalability, reduced load on backend systems, and fault tolerance. We discussed the drawbacks and challenges of distributed caching, such as consistency challenges, cache invalidation complexity, increased complexity and configuration, and network overhead.

Then we started tackling the problem of actually designing a distributed cache, starting with documenting the functional and non-functional requirements and then thinking about the core data structure for caches. We iteratively enhanced the solution by tackling the challenges faced. We then evaluated our design against the functional and non-functional requirements.

Lastly, we looked at two of the most popular distributed cache solutions and discussed their key features, as well as their use cases. We also discussed how scalable they are. We also discussed which considerations to look into to choose one over the other.

In the next chapter, we will explore pub/sub and distributed queues.

7

Pub/Sub and
Distributed Queues

In today's digital age, where data is generated at an unprecedented scale and speed, the need for robust, scalable, and efficient systems for processing and distributing this data is paramount. This is where distributed systems come into play, particularly through the use of messaging patterns such as **Publish-Subscribe (Pub/Sub)** systems and distributed queues.

As we have discussed in previous chapters, a distributed system is a network of independent components designed to work together to achieve a common goal. These components communicate and coordinate their actions by passing messages to one another. The two primary messaging patterns used in distributed systems are as follows:

- Pub/sub systems
- Distributed queues

Both pub/sub systems and distributed queues are crucial in the architecture of modern distributed systems, particularly in microservices architecture. They provide a means for services to communicate in a loosely coupled manner, enhancing scalability and reliability.

We will be covering the following topics in this chapter:

- The evolution of distributed systems
- Designing a distributed queue
- Designing a Pub/sub system
- Kafka
- Kinesis

The evolution of distributed systems

Distributed systems have evolved from monolithic architectures, where components are tightly coupled and reside in a single service, to microservices architectures, where services are decoupled and communicated via message passing. A distributed queue is a queue that is spread across multiple servers. It allows for the storage and forwarding of messages. Queues are typically used in scenarios where it's crucial to process messages in the order they were sent or where each message is intended for a single consumer. The upcoming sections will delve deeper into designing and implementing distributed queues and pub/sub systems.

Designing a distributed queue

Distributed queues are a fundamental part of modern distributed systems. They provide a reliable way to process and distribute messages across different parts of a system, ensuring that the load is balanced and that the system can scale with demand. Let us now learn about the advantages of leveraging distributed queues.

The advantages of using queues

Queues help with load management, allowing different parts of the system to independently scale and operate and provide reliability:

- **Load management**: Queues help in managing workloads by acting as a buffer between message producers and consumers. This buffering allows for handling bursts of messages without overwhelming the system.

- **Decoupling system components**: Queues enable different parts of a system to operate independently. Producers and consumers do not need to be aware of each other's states, leading to a more modular and maintainable architecture.

- **Reliability and consistency**: By using queues, systems can ensure that messages are processed in a fail-safe manner. If a consumer fails to process a message, the queue can redeliver it, ensuring data consistency.

Now that we have seen the benefits of using queues, let us look at some considerations when designing a queue system.

Designing and implementing a distributed queue

Designing a distributed queue involves several key considerations around load management, reliability, and providing delivery guarantees:

- **Scalability**: Determining how the queue will handle increasing loads and message sizes

- **Fault tolerance**: Ensuring that the queue remains operational even if some components fail

- **Message integrity**: Guaranteeing that messages are delivered exactly once and in the correct order, whenever required

Let us now discuss the key features of a distributed queue.

Key features of a distributed queue

A distributed queue consists of several integral components that work together to manage the flow of messages efficiently. Understanding their key features is essential for designing a robust distributed queue system:

- **Queue manager**: The queue manager is the core component that handles the distribution of messages within the queue. It is responsible for maintaining the order of messages, ensuring their delivery to the appropriate consumers, and managing retries in case of processing failures.

- **Message storage**: This component is where the messages are stored until they are processed by a consumer. The storage system needs to be highly reliable and capable of handling large volumes of data. It should also support fast read and write operations to ensure efficient message handling.

- **Load balancing**: Effective load balancing is crucial in a distributed queue to ensure that no single consumer is overwhelmed with messages. This involves evenly distributing messages across multiple consumers based on their current load and processing capacity.

- **Fault tolerance and recovery**: A distributed queue must be resilient to failures. This involves using mechanisms for detecting failed messages and processing attempts, as well as rerouting or retrying message delivery to ensure reliability.

- **Scalability**: The ability to scale the queue system based on the load is essential. This means being able to handle an increasing number of messages and consumers without degradation in performance.

Figure 7.1 shows how the components of the distributed queue system fit together.

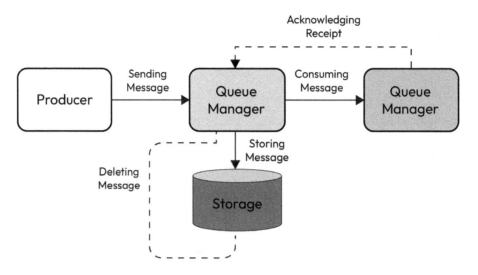

Figure 7.1: Architecture of a distributed queue system

Now that we have a brief idea about the several components that make up a distributed queue system, let us focus on some architectural considerations.

Architectural considerations

In the complex and evolving landscape of system design, the construction of a distributed queue system requires careful consideration of several architectural elements. Each aspect plays a critical role in ensuring the system's efficiency, scalability, and reliability:

- **Message ordering**: A fundamental consideration in the design process is message ordering and delivery guarantees. Depending on the application's needs, the queue may need to enforce strict message ordering or provide varying levels of delivery guarantees, such as *at-least-once* or *exactly-once* delivery. This impacts how the system manages and prioritizes the flow of data, ensuring accurate and reliable communication between different components.

- **Persistence**: Equally important is the aspect of data persistence. The design must determine the extent to which messages are stored in the system. Some applications necessitate the retention of messages until they are acknowledged by the consumer, while others may allow for a level of message loss, balancing data integrity and system performance.

- **Security**: Security and access control are also critical. Implementing robust security measures is essential to protect the integrity of messages and maintain a secure environment. This involves not only safeguarding the data itself but also controlling who has access to the queue, thus ensuring data privacy and compliance with regulatory standards.

- **Choice of technology stack**: The next phase involves the selection of the appropriate technology stack. This decision is influenced by factors such as language support, performance requirements, and compatibility with existing infrastructure. Technologies such as RabbitMQ, Apache Kafka, and Amazon SQS are popular choices, each offering unique features and capabilities suited to different types of applications.

- **Message format design**: Designing the message format is another critical aspect. This includes determining the size of messages and the methods for serialization and deserialization, as well as incorporating metadata for tracking and debugging. The format chosen can significantly affect the performance and ease of use of the queue.

- **Fault tolerance**: Ensuring fault tolerance is crucial for system reliability. Strategies for dealing with component failures, such as retry mechanisms and dead letter queues for unprocessable messages, are necessary for maintaining continuous operation and minimizing data loss.

In conclusion, designing a distributed queue is a multifaceted process that requires a careful balance of technical and practical considerations. Each element, from message ordering to security measures, plays a vital role in creating a robust and efficient system that meets the demands of modern applications.

Now that we have discussed the basics of distributed queues, let us understand a Pub/Sub system that is used extensively in the real world.

Designing a pub/sub system

The pub/sub model is a messaging pattern whereby messages are published by producers (publishers) on a specific topic and consumers (subscribers) receive messages based on their subscription to those topics. This model is highly effective for broadcasting information to multiple consumers and is widely used in real-time data processing systems.

In a pub/sub system, topics are categories or channels to which messages are published by producers to which consumers can subscribe to receive messages. Topics provide a logical grouping of related messages, allowing for targeted distribution of information to interested parties.

Let us look at some of the key characteristics of the pub/sub system.

Key characteristics of pub/sub systems

A good pub/sub system should allow independent scaling of producers and consumers while providing scale and reliability guarantees. We will now touch upon these key characteristics.

- **Decoupling of producers and consumers**: In a pub/sub model, publishers and subscribers are loosely coupled. Publishers do not need to know about the subscribers, which allows for greater flexibility and scalability.

- **Scalability and efficiency**: The pub/sub model can handle a large number of messages and distribute them to many subscribers efficiently, making it suitable for large-scale distributed systems.

- **Flexibility in message processing**: Subscribers can process messages in various ways, allowing for diverse applications ranging from real-time data analytics to event-driven architectures.

Pub/sub systems need to inherently support scalability, reliability, and decoupled architecture. In order to design an effective system, several critical architectural decisions must therefore be made around managing topics, routing messages, managing subscribers, and defining the **Quality of Service (QoS)** for the system:

- **Topic management**: It's crucial to define how topics are created, managed, and destroyed. This includes considering how to handle dynamic topic creation and deletion. Topics need to be created to provide a destination for publishers to send messages and for subscribers to receive messages from. This includes considering how to handle dynamic topic creation and deletion based on the evolving needs of the system and its users.

- **Message routing**: Developing a mechanism for routing messages from publishers to the appropriate subscribers is essential.

- **Subscriber management**: Handling subscriber registration, maintaining the list of active subscribers, and managing their subscriptions is necessary. Managing subscriptions involves allowing subscribers to specify which topics they want to receive messages from and handling the process of subscribing and unsubscribing from topics based on their changing interests or requirements.

- **QoS**: Deciding on the levels of service, such as message delivery guarantees (*at-most-once, at-least-once, exactly-once*), and handling message ordering is also vital.

- **Scalability and load balancing**: Ensuring that the system can scale to accommodate a growing number of publishers and subscribers, including strategies for load balancing, is also highly important.

Figure 7.2 shows a typical pub/sub system.

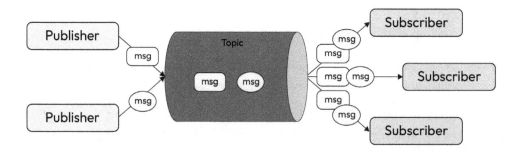

Figure 7.2: Architecture of a pub/sub system with publishers, subscribers, topics, and messages

In the preceding diagram, there are multiple publishers publishing messages on the same topic and multiple consumers consuming these messages.

Let us now learn about some of the design considerations in building a pub/sub system.

Pub/sub system design considerations

The pub/sub model is particularly relevant in the context of microservices architecture. It enables services to communicate asynchronously and react to events, leading to highly responsive and flexible systems. We will explore this in more detail, providing insights into how pub/sub systems can be effectively integrated into microservices.

When architecting a pub/sub system, two pivotal factors that warrant meticulous attention are scalability and reliability. These elements are the cornerstones of ensuring that the system not only accommodates high volumes of messages seamlessly but also maintains unwavering performance regardless of varying operational loads.

Scalability is the system's ability to gracefully handle increases in workload without compromising performance. In a pub/sub context, this translates to effectively managing an escalating number of messages and subscribers. Let us look at scalability aspects of pub/sub systems:

- **Horizontal versus vertical scaling**: This critical decision determines the scaling strategy. Horizontal scaling, which involves adding more machines to the system, offers enhanced flexibility and robustness. It allows the system to expand in capacity with the addition of resources, aligning with the fluctuating demands of message throughput. On the other hand, vertical scaling, which entails bolstering existing machines with more resources, might offer a simpler short-term solution but often lacks the long-term scalability and resilience of horizontal scaling. We have previously discussed horizontal and vertical scaling in *Chapter 2*

- **Dynamic load balancing**: In pub/sub systems, load balancing is typically handled by message brokers to distribute the load across subscribers within a subscriber group. The aim is to decouple the concerns of production and consumption load distribution from publishers and subscribers as much as possible, minimizing the need for message orchestration patterns. By distributing the load across multiple brokers, the system can achieve increased scalability and stability. Brokers dynamically adapt to changes in message volume and subscriber demand, ensuring that no single node becomes a bottleneck and maintaining optimal performance.

- **Topic partitioning**: This technique involves dividing topics into smaller, more manageable partitions. In a pub/sub system, topics are logical channels or categories used to organize and distribute messages. Publishers send messages to specific topics, while subscribers express interest in one or more topics and receive messages published to those topics. Topics provide a way to decouple the production and consumption of messages, allowing multiple publishers to send messages to the same topic and multiple subscribers to receive messages from the same topic independently. This decoupling enables greater scalability and flexibility in the system, as publishers and subscribers can be added or removed without affecting each other. Topics can be created dynamically based on the needs of the system and can have varying levels of granularity, from broad categories to highly specific subjects, depending on application requirements. This division not only facilitates load distribution across the system but also enhances throughput. By spreading the workload across multiple nodes, topic partitioning plays a significant role in optimizing the performance of a pub/sub system.

Reliability in a pub/sub system is pivotal for ensuring consistent operation and trustworthiness, particularly in handling message delivery and system resilience. Here are the design decisions to consider for reliability of pub/sub systems:

- **Message delivery guarantees**: Implementing varying levels of delivery guarantees is fundamental to ensuring message integrity. Options such *as at-most-once, at-least-once*, or *exactly-once* delivery provide different assurances depending on the criticality of the message content. These guarantees are instrumental in defining how the system handles potential data loss or duplication, thus impacting the reliability that is perceived by end users.

- **Fault tolerance**: A robust pub/sub system must be resilient to failures. Designing for fault tolerance involves strategies such as replicating data across different nodes and implementing message re-delivery mechanisms. This ensures that the system remains operational and efficient even in the face of component failures, thereby maintaining uninterrupted service.

- **Message ordering**: For applications where the sequence of messages is crucial, ensuring correct message order is a key aspect of reliability. This involves designing the system to maintain the chronological integrity of messages, which can be especially challenging in distributed environments where messages may traverse various paths.

In essence, these high-level considerations of scalability and reliability form the backbone of a well-designed pub/sub system. They are crucial in ensuring that the system not only meets the current demands but is also well-prepared for future expansions and challenges. This focus helps in creating a resilient, efficient, and future-proof messaging infrastructure.

Given that we have discussed the importance of scalability and reliability in pub/sub systems, let us now delve into some aspects of their architecture.

Subscriber management and message routing

Managing subscribers and efficiently routing messages to them are vital components of a pub/sub system. This involves the following:

- **Subscriber registration and management**: Handling the process by which subscribers can register, deregister, and manage their topic subscriptions

- **Efficient message routing**: Developing algorithms to ensure that messages are routed to the right subscribers in the most efficient way, minimizing latency and resource usage

QoS levels

Different applications require different levels of QoS. For instance, some systems might need guaranteed delivery (*at-least-once* or *exactly-once*), while others might prioritize lower latency over delivery guarantees.

Let us discuss some steps that you, as a system design architect, can take to architect pub/sub systems in the real world.

Building a basic pub/sub system

The practical implementation of a pub/sub system involves several key steps, which cover everything from initial setup to ensuring efficient and robust operation. Here, we will outline the process of creating a basic pub/sub system:

1. **Choosing the right tools and technologies**:

   - Select a pub/sub platform (such as Apache Kafka, RabbitMQ, or Google Pub/Sub) that fits the requirements of your system in terms of scalability, reliability, and feature set.

   - Decide on the programming languages and frameworks that will be used for developing publishers and subscribers.

2. **Setting up the pub/sub infrastructure**:

   - Establish topics for publishers to send messages to.

   - Configure the pub/sub system, including setting up the necessary infrastructure for message storage, routing, and processing.

3. **Developing publishers and subscribers**:

   - Write the code for publishers to send messages to topics.

   - Create subscribers that listen to topics and process incoming messages.

   - Implement error handling and retry mechanisms for reliable message processing.

4. **Testing and optimization**:

   - Conduct thorough testing to ensure that the system works as expected under different scenarios.

   - Optimize performance, including tweaking configurations for load balancing and message routing.

Now that we have extensively covered pub/sub system design, let us look at some real-world systems in the next section

Kafka

Apache Kafka is a distributed streaming platform that has become synonymous with handling high-throughput, real-time data feeds. Kafka is designed to be durable, fast, and scalable, making it an ideal choice for implementing both pub/sub and queue-based messaging patterns. Before we get into the details, let us recap some core concepts of Kafka:

- **Topics**: In Kafka, records are published to a category called *topics*, to which many consumers subscribe.

- **Producers and consumers**: In Kafka, producers publish the data. Consumers read that data. This data is published and read from topics. Kafka's power lies in allowing producers and consumers to scale independently in a decoupled manner.

- **Brokers**: Kafka clusters consist of one or more servers known as brokers, which are responsible for maintaining published data, managing subscribers, tracking consumption offsets, and fanning out data to consumers on demand.

- **Partitions and replication**: Each topic can be split into partitions, which allows for data to be spread over multiple brokers for fault tolerance and increased throughput.

In practice, as producers send messages to a topic, they assign these messages to different partitions within the topic. Consumers then subscribe to these topics and read messages from the partitions. Kafka takes care of maintaining the offset of messages for each consumer, ensuring that each message is read once and in order. The use of replication across multiple brokers protects against data loss and contributes to the overall reliability of the system.

Overall, Kafka's architecture, characterized by its use of topics, partitions, and brokers, is adept at balancing loads by evenly distributing messages and consumer requests. This not only ensures scalability but also maintains a high level of performance, making Kafka an ideal platform for handling real-time data feeds in distributed environments.

The following figure demonstrates the basic architecture of Kafka. As you can see in the figure, messages are published to topics that contain partitions. Brokers are hubs that manage topics and partitions. ZooKeeper is used to determine which broker is the leader for each partition and as a service registry for all the brokers.

> **Note:**
> In recent versions of Kafka (starting from version 2.8), the dependency on ZooKeeper has been reduced. Kafka now uses its own internal components, such as the Kafka Controller Broker and the Kafka Raft consensus protocol, to manage cluster metadata and coordinate brokers. However, some older versions of Kafka may still rely on ZooKeeper for these purposes.

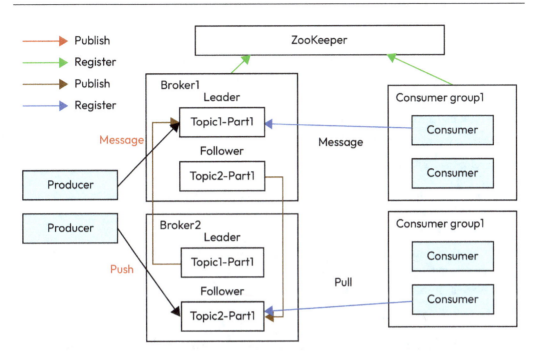

Figure 7.3: Kafka architecture

Now that we have a better understanding of the Kafka architecture, let us focus on understanding the importance of Kafka in distributed systems and what steps need to be followed to deploy it.

The importance of Kafka in distributed systems

Kafka is not just a messaging system but a full-fledged event streaming platform. It plays a crucial role in modern distributed systems, particularly in scenarios that require handling large volumes of data with minimal latency. Deploying Kafka in a distributed environment involves several crucial steps, from initial setup to fine-tuning for optimal performance:

1. **Setting up a Kafka cluster:**

   - **Cluster configuration:** Configuring a Kafka cluster involves setting up multiple brokers to ensure fault tolerance and high availability. This includes deciding on the number of brokers and their configuration for optimal load distribution.

   - **ZooKeeper integration:** Kafka uses ZooKeeper for managing cluster metadata and coordinating brokers. Setting up and configuring ZooKeeper is an essential step in deploying Kafka.

2. **Creating and configuring topics**:

    - **Topic creation**: Topics in Kafka are where messages are stored and published. Creating topics with the right configurations, such as the number of partitions and the replication factor, is crucial for performance and reliability.

    - **Topic management**: Understanding how to manage topics, including modifying configurations, deleting topics, and monitoring topic performance, is crucial.

3. **Developing Kafka producers and consumers**:

    - **Writing producers**: Implementing producers that publish messages to Kafka topics, including handling serialization of messages and managing connections to the Kafka cluster, is an essential step.

    - **Creating consumers**: It's important to develop consumers that subscribe to topics and process messages, with a focus on concurrency and offset management, as well as on ensuring message processing reliability.

4. **Monitoring and maintenance**:

    - **Cluster monitoring**: Monitoring is a crucial aspect for any distributed system deployment at scale, and therefore, a successful deployment of Kafka should include tools and techniques for monitoring the cluster health and performance metrics to identify potential issues.

    - **Performance tuning**: Ensure that you observe the best practices for tuning Kafka performance, such as optimizing producer and consumer configurations, and tuning network and disk IO.

We have talked about the architecture of Apache Kafka. Now, let us look at Kafka Streams, which provides a stream processing library on top of Kafka.

Kafka Streams

Kafka Streams is a powerful stream-processing library built on top of Apache Kafka. It allows developers to build scalable, fault-tolerant, and real-time stream processing applications using a high-level **Domain-Specific Language (DSL)** or a low-level Processor API. Kafka Streams enables you to process and analyze data directly within Kafka, eliminating the need for separate stream processing frameworks. The following are the key features of Kafka Streams:

- **Native Kafka integration**: Kafka Streams is tightly integrated with Kafka, allowing seamless reading from and writing to Kafka topics. It leverages Kafka's features, such as partitioning and replication, for fault tolerance and scalability.

- **High-level DSL**: Kafka Streams provides a high-level DSL that simplifies the development of stream processing applications. The DSL includes operations such as map, `filter`, `join`, and `aggregate`, which can be combined to express complex processing logic declaratively.

- **Low-level Processor API**: For more advanced use cases, Kafka Streams offers a low-level Processor API that gives developers fine-grained control over the processing topology. The Processor API allows for custom processing logic and state management.

- **Stateful processing**: Kafka Streams supports stateful processing, enabling you to build applications that maintain and update state over time. It provides state stores, which are fault-tolerant and can be used to store and query application state.

- **Exactly-once processing**: Kafka Streams guarantees exactly-once processing semantics, ensuring that each record is processed exactly once, even in the presence of failures. This is achieved through the use of transaction support in Kafka.

- **Scalability and fault tolerance**: Kafka Streams applications can be scaled horizontally by adding more instances, with Kafka handling the distribution of partitions among the instances. It also provides fault tolerance through the use of Kafka's replication and failover mechanisms.

Stream processing topology

In Kafka Streams, a stream processing application is defined as a topology. A topology is a graph of stream processors (nodes) and streams (edges). Each processor performs a specific operation on the input data and produces an output stream. The topology defines how data flows through the processors and how the processors are connected.

Kafka Streams provides a fluent API for building topologies using the high-level DSL. You can define sources (input topics), processors (transformation operations), and sinks (output topics) to create a complete stream processing pipeline.

Example use cases

Kafka Streams is suitable for a wide range of stream processing scenarios, such as the following:

- **Real-time data analytics**: Analyzing streaming data in real-time to derive insights, detect anomalies, or generate metrics

- **Event-driven architectures**: Building event-driven systems that react to and process events in real-time, enabling communication and coordination between microservices

- **Data enrichment and transformation**: Enriching streaming data with additional information from external sources or transforming data into a desired format for downstream consumption

- **Fraud detection**: Analyzing streaming transactions or user behavior in real time to identify and prevent fraudulent activities

- **Streaming ETL**: Performing real-time **Extract, Transform, and Load** (ETL) operations on streaming data to enable continuous data integration and processing

By leveraging Kafka Streams, you can build powerful and scalable stream processing applications that seamlessly integrate with your Kafka-based architecture. Its combination of high-level abstractions, low-level control, and native Kafka integration makes it a compelling choice for real-time data processing needs.

In the next section, we will discuss another significant player in the world of real-time data streaming and processing: **Amazon Kinesis**.

While Kafka offers a powerful solution for handling high-throughput data feeds, the dynamic and ever-expanding field of distributed systems continually demands diverse approaches and technologies. Kinesis represents another facet of this evolving landscape, providing a fully managed, cloud-based service that addresses specific challenges and use cases in streaming data. As we transition from Kafka to Kinesis, we'll explore how Kinesis complements and differs from Kafka, particularly in its approach to stream processing in cloud environments, and how it fits into the broader picture of modern data-driven applications.

Kinesis

Kinesis is a cloud-based service from **Amazon Web Services** (**AWS**) that facilitates real-time processing of large data streams. Kinesis is designed to handle massive amounts of data with very low latencies, making it an ideal solution for real-time analytics, log and event data processing, and large-scale application monitoring. Kinesis comprises several core components, each serving distinct roles in the stream processing ecosystem:

- **Kinesis Data Streams** form the backbone of the Kinesis service. They enable the capture and processing of large data records in real time. Data producers, ranging from log generators to real-time event sources, continuously feed data into these streams. The flexibility of Kinesis Data Streams lies in their ability to handle high-throughput ingestion from multiple sources simultaneously, making them ideal for scenarios requiring immediate data processing and analysis.

- **Kinesis Data Firehose** complements Kinesis Data Streams by facilitating the direct loading of streaming data into other AWS services and external data stores. This service captures the data from streams, optionally transforms it, and then automatically loads it into repositories such as Amazon S3, Amazon Redshift, Amazon Elasticsearch, or even Splunk. The seamless integration of Kinesis Data Firehose with these data stores simplifies the architecture for real-time analytics applications, allowing for efficient data storage, querying, and visualization.

- Lastly, **Kinesis Data Analytics** offers a powerful solution for analyzing streaming data using standard SQL queries. This component allows users to write SQL queries to process data directly within the stream, enabling real-time analytics and decision-making. Kinesis Data Analytics is particularly powerful when combined with Kinesis Data Streams, as it allows for the immediate analysis of incoming data without the need for batch processing or data offloading.

Together, these components create a comprehensive ecosystem for real-time data processing in the cloud. From data ingestion and storage to processing and analysis, Kinesis offers scalable and efficient solutions, making it a go-to choice for applications that require real-time insights from streaming data.

Summary

In this chapter, we embarked on a detailed journey through the realms of pub/sub and distributed queue systems, which are integral to the architecture of modern distributed systems. We delved into the intricacies of designing and implementing distributed queues, focusing on their scalability, reliability, and message integrity. The chapter also explored pub/sub systems, emphasizing their role in decoupling components and enhancing system flexibility. We covered practical aspects, including integration with microservices. Finally, we examined Apache Kafka and Amazon Kinesis, shedding light on their core concepts and operational dynamics. This chapter has armed you with a comprehensive understanding of these crucial components, preparing you to design and implement efficient, scalable distributed systems.

In the next chapter, we will cover some cornerstones of a well-architected distributed system such as API design, security, and metrics.

Part 3: System Design in Practice

In this *Part*, we transition from theoretical concepts to practical application, exploring how the principles and components we've discussed are used to design real-world systems. This section is designed to bridge the gap between theory and practice, providing you with hands-on experience in tackling complex system design challenges.

By the end of this *Part*, you'll have a robust toolkit for tackling any system design challenge, whether in a professional setting or an interview scenario.

This section contains the following chapters:

- *Chapter 8, Design and Implementation of System Components: API, Security, and Metrics*
- *Chapter 9, System Design – URL Shortener*
- *Chapter 10, System Design – Proximity Service*
- *Chapter 11, Designing a Service Like Twitter*
- *Chapter 12, Designing a Service Like Instagram*
- *Chapter 13, Designing a Service Like Google Docs*
- *Chapter 14, Designing a Service Like Netflix*
- *Chapter 15, Tips for Interviewees*
- *Chapter 16, System Design Cheat Sheet*

8

Design and Implementation of System Components: API, Security, and Metrics

In the realm of software engineering, the design and implementation of system components play a pivotal role in determining the efficiency, reliability, and scalability of a system. This chapter delves into the intricacies of key system components, namely APIs (with a focus on REST and gRPC), API security, logging, metrics, alerting, and tracing in a distributed system. The objective is to provide a comprehensive understanding of these components, their design principles, and their contribution to the overall performance of a system.

Application Programming Interfaces (APIs) serve as the communication bridge between different software components. They have become increasingly important in today's world of microservices and distributed systems. This chapter will provide an in-depth exploration of REST and gRPC APIs, two popular approaches to building APIs, each with its unique strengths and suitable use cases.

API security is another critical aspect that will be covered in this chapter. As APIs are often targeted by attackers, understanding and implementing robust API security measures is crucial. We will delve into the basics of API security, including authentication, authorization, secure communication, and rate limiting.

Logging and metrics are essential tools for monitoring and debugging in a distributed system. They provide insights into the system's behavior and performance, helping engineers identify and resolve issues. This chapter will guide you through the design and implementation of distributed logging and the relevance of metrics.

Alerting is a proactive way to manage system incidents. It notifies engineers about potential issues before they escalate into major problems. This chapter will discuss the importance of alerting, how to set up effective alerts, and how to respond to them.

Finally, we will explore tracing, a technique used to track a request as it travels through various services in a distributed system. Tracing helps in diagnosing performance issues and understanding the system's behavior.

Here is a list of topics covered in this chapter:

- REST APIs

- gRPC APIs

- Comparing REST and gRPC

- API security

- Distributed logging

- Metrics in a distributed system

- Alerting in a distributed system

- Tracing in a distributed system

By the end of this chapter, you will have gained practical skills in designing and implementing these system components, which are essential for building robust, scalable, and efficient systems. Let's embark on this journey by exploring REST and gRPC APIs.

REST APIs

Representational State Transfer (**REST**) is an architectural style for designing networked applications. It has gained popularity due to its simplicity and the use of standard HTTP methods, which are understood by most developers. Let us learn about the design principles, use cases, and strengths and weaknesses of REST APIs.

Design principles of REST APIs

REST APIs are built around resources, which are any kind of object, data, or service that can be accessed by the client. A resource is identified by a **Uniform Resource Identifier** (**URI**), and the API interacts with these resources using HTTP methods such as GET, POST, PUT, DELETE, and others. These methods correspond to **create**, **read**, **update**, and **delete** (**CRUD**) operations in database systems.

A key feature of REST is its statelessness. Each request from a client to a server must contain all the information needed to understand and process the request. This makes the server's operations more predictable and reliable as it doesn't need to retain any context between requests. Let us now look at the use cases of REST APIs.

Use cases for REST APIs

REST APIs are particularly suitable for public APIs exposed over the internet. Their use of HTTP protocol makes them easily consumable by any client that can send HTTP requests. This includes web browsers, mobile applications, and other servers.

REST APIs are also a good fit for CRUD-based operations where the system is primarily dealing with entities that need to be created, read, updated, or deleted. For example, a web application that manages a list of users or products can benefit from a RESTful API design. Let us look at the strengths and weaknesses of REST APIs.

Strengths and weaknesses

REST APIs are simple to understand and use. They leverage HTTP's built-in methods and status codes, making them intuitive for developers familiar with HTTP. However, REST APIs can be less efficient for complex operations that require multiple requests to complete. They also use JSON for data exchange, which can be verbose and lead to larger payloads.

In the next section, we will explore gRPC APIs, a different approach to building APIs that can address some of the limitations of REST. By understanding both REST and gRPC, you will be better equipped to choose the right approach for your specific use case.

gRPC APIs

Google Remote Procedure Call (gRPC) is a high-performance, open-source framework for executing remote procedure calls. It was developed by Google and is based on the HTTP/2 protocol. Unlike REST, which is data-centric, gRPC is function-centric, making it a powerful tool for creating highly efficient APIs. We will now discuss the design principles, use cases, and strengths and weaknesses of gRPC APIs.

Design principles of gRPC APIs

gRPC uses **Protocol Buffers (protobuf)** as its interface definition language. Protobuf is a language-neutral, platform-neutral, extensible mechanism for serializing structured data. It's more efficient and faster than JSON, which is commonly used in REST APIs.

gRPC allows you to define services in a `.proto` file, and then automatically generates client and server stubs in a variety of languages. This makes it easier to create and maintain APIs, as changes to the service definition are automatically propagated to the client and server code.

One of the main advantages of gRPC is its support for multiple programming languages, making it a good choice for polyglot environments. It also supports features including authentication, load balancing, and bidirectional streaming. Let us now look at some of the use cases.

Use cases for gRPC APIs

gRPC is particularly suitable for microservices architectures where services need to communicate with each other frequently and efficiently. Its support for bidirectional streaming and its small message size due to Protocol Buffers make it a good choice for real-time applications.

gRPC is also a good fit for systems that require high-performance inter-service communication, as it reduces the overhead of communication between services. Let us talk about the strengths and weaknesses of gRPC APIs.

Strengths and weaknesses

gRPC APIs are highly efficient and versatile, supporting a wide range of use cases. They offer significant performance benefits over REST APIs, especially in terms of payload size and speed. gRPC has strictly typed data definition contracts guaranteed by protobufs. However, gRPC APIs can be more complex to set up and debug due to their binary format and the need for HTTP/2 support. Changes to gRPC communication can require data definition changes to be made and deployed as code changes, thus increasing the change release time. On the other hand, changing messages and testing these changes is quicker when using REST v/s gRPC.

We will now compare REST and gRPC, discussing when to use each based on specific requirements. This comparison will provide a deeper understanding of these two approaches, enabling you to make an informed decision for your system design.

Comparing REST and gRPC

REST and gRPC are two powerful approaches to building APIs, each with its unique strengths and suitable use cases. Understanding the differences between them can help you make an informed decision based on your specific requirements. We will compare them across several dimensions including performance, ease of use, compatibility, and streaming support. *Figure 8.1* shows typical request and response structures for both REST and gRPC. REST endpoints are easier to test and are generally preferred for public APIs, whereas gRPC APIs are preferred for inter-service communications and have a binary format.

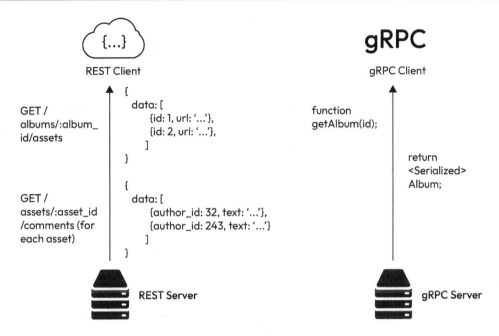

Figure 8.1: REST and gRPC APIs

While both gRPC and REST APIs excel at building efficient communication protocols for services, a key difference lies in their browser compatibility. gRPC, designed for machine-to-machine communication, is not directly supported by web browsers due to limitations in browser APIs. This makes REST APIs the preferred choice for browser-based applications where data needs to be exchanged between the client-side (web browser) and the server. However, for service-to-service communication where efficiency and performance are paramount, gRPC remains a powerful option. We will now discuss the differences between the two APIs based on factors such as performance, ease of use, compatibility, streaming support, and use cases:

- **Performance**: gRPC generally has better performance than REST due to its use of HTTP/2 and Protocol Buffers. HTTP/2 can send multiple requests in parallel over a single TCP connection, reducing the latency inherent in HTTP/1.1. Protocol Buffers are a more efficient data format than JSON, leading to smaller payloads.

- **Ease of use**: REST APIs are simpler to use and understand, especially for developers already familiar with HTTP methods and status codes. They can easily be tested and debugged using standard tools such as cURL or Postman. On the other hand, gRPC APIs require specific tools for testing and debugging due to their binary format.

- **Compatibility**: REST APIs have broader compatibility as they use HTTP, which is universally supported by all internet-connected devices. gRPC, however, requires HTTP/2 support, which might not be available on all platforms or network infrastructures.

- **Streaming support**: gRPC supports bidirectional streaming, allowing both the client and server to send data independently of each other. REST, however, only supports request-response communication.

- **Use cases**: REST is a good choice for public APIs exposed over the internet, especially for CRUD-based operations. It's also suitable when you need broad compatibility across different platforms and network environments.

gRPC is a better choice for high-performance inter-service communication, especially in a microservices architecture. It's also suitable for real-time applications due to its support for bidirectional streaming.

In conclusion, the choice between REST and gRPC should be based on your specific use case, considering factors such as performance requirements, ease of use, compatibility, and the type of communication needed.

In the next section, we will delve into API security, a critical aspect of any API design.

API security

API security is a critical aspect of any system design, especially in today's world where data breaches are common. As APIs serve as the communication bridge between different software components, they are often targeted by attackers. Therefore, understanding and implementing robust API security measures is crucial. We will discuss authentication and authorization as it pertains to API security in the next sections. *Figure 8.2* shows the difference between authentication and authorization. Authentication is concerned with answering the question, *"Who are you?"*, which is akin to logging into a website with your username and password, whereas authorization deals with checking users' permissions to access data, thus answering the question, *"Are you allowed to do that?"* Let us now delve into explaining both authentication and authorization.

Figure 8.2: Authentication vs. Authorization

Authentication

Authentication is the process of verifying the identity of a user, system, or application. It's the first line of defense in API security. There are several methods to implement authentication in APIs, each with its strengths and weaknesses.

- **API keys**: API keys are a simple method where the client sends a key in the header of the HTTP request. However, they are not the most secure method as they can be easily intercepted and don't provide fine-grained access control.

- **OAuth**: OAuth is a more secure method that allows users to grant limited access to their resources from one site to another site, without having to expose their credentials. OAuth 2.0, the latest version, is widely used in the industry and supports different "flows" for web applications, desktop applications, mobile applications, and smart devices.

- **JSON Web Tokens (JWT)**: JWT is a streamlined method for securely exchanging information between two entities. It encapsulates claims in a JSON format, which are then encapsulated within a **JSON Web Signature** (**JWS**) structure. This design allows for the digital signing or integrity protection of the claims using a **Message Authentication Code** (**MAC**), and optionally, encryption. The compact and URL-safe nature of JWTs makes them an ideal choice for transmitting data across different systems or networks.

We have thus seen the different ways in which authentication is implemented. Authentication helps to answer the question *"Who are you?"*, i.e., it validates that the system is being accessed by the right person. Now, let us understand authorization, which checks for the access permissions of a given user.

Authorization

Once a user is authenticated, the next step is to ensure they have the correct permissions to access the resources they are requesting. This is known as authorization. Let us review some ways in which authorization is typically implemented in large-scale systems:

- **Role-Based Access Control** (**RBAC**): RBAC is a popular method for implementing authorization, where permissions are associated with roles, and users are assigned roles. This simplifies managing permissions as you only need to manage roles, not individual user permissions.

- **Attribute-Based Access Control** (**ABAC**): ABAC, on the other hand, defines permissions based on attributes. These attributes can be associated with a user, a resource, an environment, or a combination of these. ABAC provides more fine-grained control than RBAC but can be more complex to implement.

These are two basic ways that authorization has been implemented in many large-scale systems. To understand API security, we also need to understand that APIs typically require secure communications and rate limiting, which we define below.

Secure communication of APIs

When building APIs, it's crucial to prioritize security to safeguard data in transit. Using **Hypertext Transfer Protocol Secure** (**HTTPS**) is a fundamental practice in achieving this. HTTPS ensures that the communication between the client and the server is encrypted, making it difficult for potential eavesdroppers or attackers to intercept and decipher the transmitted data. This is especially important when dealing with sensitive information such as passwords, personal details, or financial transactions. Additionally, maintaining up-to-date **Transport Layer Security** (**TLS**) configurations is vital. As new vulnerabilities are discovered and security standards evolve, keeping your TLS settings current helps protect against emerging threats and ensures compliance with industry best practices for encryption and data integrity.

Rate limiting

Rate limiting is an essential mechanism for API management and security. By restricting the number of requests a client can make within a specified timeframe, rate limiting helps prevent abuse and overuse of your API. This is particularly important for defending against **Denial of Service** (**DoS**) attacks, where an attacker attempts to overwhelm your server by flooding it with an excessive number of requests, leading to service degradation or downtime. Implementing rate limiting also aids in managing server resources and ensuring fair usage among clients. By setting reasonable limits, you can help ensure that no single user or application monopolizes your API, allowing for a more equitable distribution of resources and a better overall experience for all users.

There are several popular rate-limiting algorithms you can choose from, each with its own advantages and disadvantages. Here are a few common ones:

- **Token bucket algorithm**: This algorithm visualizes rate limits as a bucket with a fixed number of tokens. Each request consumes a token, and new tokens are added to the bucket at a set rate. Requests are rejected if there are no available tokens.

- **Leaky bucket algorithm**: Similar to the token bucket, the leaky bucket algorithm has a bucket but with a hole at the bottom. Tokens are added at a set rate, but they also leak out at a constant rate. Requests are rejected if the bucket is full.

- **Sliding window algorithm**: This algorithm tracks the number of requests made within a specific time window. If the request count exceeds the limit during that window, subsequent requests are throttled.

The best algorithm for your specific needs depends on factors such as desired granularity, burstiness tolerance, and implementation complexity.

We have discussed the basics of API security in this section. In the following section, we will transition into the importance of logging and metrics in a distributed system.

Distributed systems logging

In a distributed system, where multiple services are running on different machines, logging becomes a critical aspect of system monitoring and debugging.

Logging is the process of recording events in a system. It provides a way to understand what's happening inside your application, helping you identify patterns, detect anomalies, and troubleshoot issues. In a distributed system, logs from different services can provide a holistic view of the system's behavior, making it easier to identify and resolve issues. Most large-scale systems use some form of centralized logging infrastructure, where logs are collected from different systems into a central logging repository. We will now delve into centralized logging.

Centralized logging

In a distributed system, logs are generated by different services running on different machines. These logs need to be collected and stored in a central location for easy access and analysis. This is known as **centralized logging**.

Benefits of centralized logging

The following are some of the benefits of centralized logging:

- **Ease of access**: All logs are available in one place, making it easier to search and analyze them
- **Correlation**: Logs from different services can be correlated based on timestamps or unique identifiers, providing a complete picture of a transaction or operation
- **Long-term storage**: Logs can be archived for long-term storage and compliance purposes

Figure 8.3 shows an example of the most commonly used centralized logging architecture, which leverages Logstash for log collection, parsing, and transformation; Elasticsearch for storing, indexing; and searching logs; and Kibana for visualizations and analysis. This is just one of the different ways in which you can build your logging infrastructure.

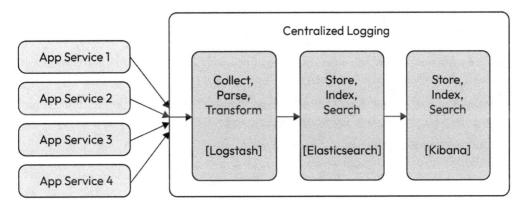

Figure 8.3: An example of a centralized logging architecture

When designing your logging strategy, consider what information to include in your logs. At a minimum, each log entry should contain the following:

- **Timestamp**: The date and time the event occurred
- **Service name**: The name of the service that generated the log
- **Severity level**: The severity of the event (e.g., INFO, WARNING, ERROR)
- **Message**: A descriptive message explaining the event
- **Context**: Additional context about the event, such as user ID, transaction ID, and so on

Building upon the importance of logging and the basics of centralized logging, let's delve into the tools and best practices for implementing distributed logging.

Open source tools for centralized logging

There are various open-source tools available for centralized logging, each offering unique features and capabilities:

- **Logstash**: A component of the Elastic Stack, Logstash is designed to efficiently process data from various sources, transform it, and then route it to a storage destination like Elasticsearch.
- **Fluentd**: Fluentd is a versatile open-source data collector that allows for the aggregation and management of data from multiple sources, facilitating easier analysis and utilization of the information.
- **Graylog**: Graylog is a prominent solution for centralized log management that adheres to open standards. It is engineered to collect, store, and enable real-time analysis of vast amounts of machine-generated data.

Let us now talk about some of the best practices for implementing logging.

Best practices for implementing distributed logging

Implementing distributed logging can be challenging, but the following best practices can guide you through the process:

- **Use a consistent log format**: A consistent log format makes it easier to search and analyze logs. This is especially important in a distributed system where logs are generated by different services.
- **Include context in your logs**: Contextual information such as user ID, transaction ID, and other relevant data can help you understand the scope and impact of an event.
- **Handle exceptions properly**: Make sure to log exceptions along with their stack traces. This will help you identify the root cause of errors.

- **Use appropriate log levels**: Using appropriate log levels helps filter logs and reduces noise. For example, use the ERROR level for events that require immediate attention, and the INFO level for events that are useful but not critical.

- **Rotate and archive logs**: To manage storage and ensure compliance, implement a strategy for rotating and archiving logs. This involves deleting old logs and storing important logs for long-term analysis.

In the next section, we will transition into the importance of metrics in a distributed system.

Metrics in a distributed system

Metrics are a numerical representation of data measured over intervals of time. They provide a quantifiable way to assess the performance and health of your system. In a distributed system, it's important to collect metrics from all your services and aggregate them in a central location.

Metrics provide insights into the behavior and performance of your system, helping you make informed decisions about scaling, performance optimization, and troubleshooting. They can help you answer questions such as the following:

- How is the system performing?

- Is the system meeting its service level objectives?

- Are there any performance bottlenecks?

- Is the system behaving as expected?

We will now talk about the types of metrics in a distributed system, some open-source tools typically used by enterprises, and some best practices for system design architects to implement metrics at scale.

Types of metrics

There are several types of metrics you can collect, each providing different insights into your system:

- **System metrics**: These include CPU usage, memory usage, disk I/O, and network I/O. They provide insights into the resource usage of your system.

- **Application metrics**: These include request rate, error rate, and response times. They provide insights into the performance and reliability of your application.

- **Business metrics**: These are specific to your application's domain, such as the number of user sign-ups, and number of orders placed. They provide insights into the business impact of your system.

Building upon the importance and types of metrics, let's delve into the tools and best practices for implementing metrics in a distributed system.

Open source tools for metrics

Several open-source tools are available for the collection, storage, and visualization of metrics, each offering unique features and strengths:

- **Prometheus**: Prometheus is a popular open-source system for monitoring and alerting. It features a multi-dimensional data model, a powerful query language called PromQL, and seamless integration with a variety of graphing and dashboarding tools. Prometheus is particularly well suited for monitoring dynamic cloud environments and microservices architectures.

- **Graphite**: Graphite is a well-established monitoring tool that specializes in storing and graphing numeric time-series data. Its web interface, Graphite-web, allows users to create and display custom graphs based on the stored data. Graphite is known for its simplicity and scalability, making it a good choice for large-scale deployments.

- **Datadog**: Although not open source, Datadog is a widely-used SaaS-based monitoring service for cloud-scale applications. It offers comprehensive monitoring capabilities for servers, databases, tools, and services, along with advanced data analytics features. Datadog's platform is designed to provide real-time insights and alerts, helping teams quickly identify and resolve issues.

Let us now discuss some of the best practices for implementing metrics.

Best practices for implementing metrics

Implementing metrics in a distributed system can be challenging, but the following best practices can guide you through the process:

- **Identify key metrics**: Not all metrics are equally important. Identify key metrics that directly impact your service level objectives and focus on them.

- **Use a consistent naming scheme**: A consistent naming scheme makes it easier to search and analyze metrics. This is especially important in a distributed system where metrics are generated by different services.

- **Monitor error rates and latencies**: These are often the first indicators of a problem. Set up alerts to notify you when these metrics cross a certain threshold.

- **Visualize your metrics**: Use dashboards to visualize your metrics. This can help you spot trends and anomalies that might not be apparent in raw data.

In the next section, we will transition into the importance of alerting in a distributed system.

Alerting in a distributed system

Alerting is a crucial aspect of maintaining the health of a distributed system. It serves as a proactive measure to identify potential issues before they escalate into significant problems, ensuring system reliability and performance.

Alerting is an essential part of any monitoring strategy. It provides real-time notifications about system anomalies, errors, or performance issues. Without an effective alerting mechanism, teams may remain unaware of critical issues until they have caused significant damage or downtime.

Alerts can be triggered based on various conditions, such as exceeding a certain threshold of error rate, response time, or resource usage. They can also be triggered based on specific events, such as a service failure or a system-wide outage.

It is important as a system design architect to design effective and actionable alerts, without overwhelming support. We will therefore discuss this topic, some open-source tools for alerting, and some best practices for designing alerts in the next few sections.

Designing effective alerts

Designing effective alerts involves striking a balance between sensitivity and specificity. Alerts should be sensitive enough to catch real issues but specific enough to avoid false alarms, which can lead to alert fatigue.

Here are some key considerations when designing alerts:

- **Severity**: Not all alerts are created equal. Classify alerts based on their severity to help prioritize responses. Critical alerts might require immediate attention, while warnings could be addressed during regular working hours.

- **Actionability**: An effective alert is one that requires action. If an alert doesn't require any action, it might not be necessary.

- **Context**: Alerts should provide enough context to help diagnose the issue. Include relevant information such as the service affected, the time the issue occurred, and any associated error messages or logs.

Building upon the importance and design principles of alerting, let's delve into the tools and best practices for implementing alerting in a distributed system.

Open-source tools for alerting

Several open-source tools are available for configuring alerts, each offering unique features and advantages:

- **Prometheus with Alertmanager**: Prometheus is a widely-used open-source monitoring and alerting toolkit. It includes a component called Alertmanager, which is responsible for processing alerts generated by Prometheus or other client applications. Alertmanager handles tasks such as deduplication, grouping, and routing of alerts to the appropriate receiver, ensuring that alerts are managed efficiently and effectively.

- **Grafana**: Grafana is a versatile open-source analytics and visualization platform. It supports a wide range of data sources and provides powerful visualization tools, including charts and graphs. Grafana also includes an alerting feature that allows users to set up alerts based on specific conditions in their data. These alerts can be configured to send notifications through various channels, such as email, Slack, or webhook, enabling timely responses to potential issues.

- **PagerDuty**: Although not open source, PagerDuty is a popular incident response platform used by IT departments and operations teams. It integrates seamlessly with a variety of monitoring tools, including both Prometheus and Grafana, to provide centralized alerting and incident management. PagerDuty can alert users through multiple channels, including phone calls, SMS, emails, and push notifications, ensuring that critical issues are addressed promptly.

Let us now discuss some of the best practices for implementing alerting.

Best practices for implementing alerting

Implementing alerting in a distributed system can be challenging, but the following best practices can guide you through the process:

- **Avoid alert fatigue**: Alert fatigue occurs when too many alerts are triggered, causing teams to ignore them. To avoid this, ensure that your alerts are actionable and adjust their thresholds to avoid false positives.

- **Test your alerts**: Regularly test your alerts to ensure they are working as expected. This can be done during chaos engineering experiments or as part of your regular testing process.

- **Document your alerting process**: Document how to respond to each type of alert. This can include troubleshooting steps, who to escalate to, and any relevant runbooks.

In the next section, we will transition into the importance of tracing in a distributed system.

Tracing in a distributed system

Tracing is a technique used to track a request as it travels through various services in a distributed system. It provides a detailed view of how a request is processed, making it an invaluable tool for diagnosing performance issues and understanding the system's behavior.

In a distributed system, a single request can involve multiple services. Understanding how this request is processed can be challenging, especially when things go wrong. This is where tracing comes in.

Tracing provides a detailed view of a request's journey through the system. It shows the interaction between services, the latency of each service, and any errors that occurred during the processing of the request. This information can help diagnose performance issues, identify bottlenecks, and understand the overall flow of requests in the system.

In the next few sections, we will discuss the importance of tracing, some open-source tools that can be used by system design practitioners for effective tracing, and some best practices to follow when designing distributed tracing.

Distributed tracing

Distributed tracing extends the concept of tracing to a distributed system. It involves tracking a request as it travels across multiple services and machines. Each step in the request's journey is recorded as a span. A collection of spans forms a trace, which represents the entire journey of the request.

Distributed tracing provides several benefits:

- **Performance optimization**: By visualizing the flow of requests, you can identify performance bottlenecks and optimize them
- **Error diagnosis**: If a request fails, you can use the trace to identify where the error occurred and what caused it
- **System understanding**: Tracing helps you understand the flow of requests in your system, which can be useful when onboarding new team members or when planning system changes

Open-source tools for distributed tracing

Several open-source tools are available for implementing distributed tracing, each offering distinct features and capabilities:

- **Jaeger**: Jaeger is a distributed tracing system that was developed by Uber Technologies and subsequently released as open-source software. It is designed to monitor and troubleshoot microservices-based distributed systems, drawing inspiration from Google's Dapper and OpenZipkin. Jaeger provides a comprehensive set of features, including distributed context propagation, transaction monitoring, root cause analysis, and performance optimization. It supports various storage backends, including Elasticsearch, Cassandra, and Kafka, for scalable trace storage.

- **Zipkin**: Zipkin is another distributed tracing system that focuses on collecting and managing timing data for troubleshooting latency issues in microservice architectures. It provides a simple and intuitive interface for visualizing trace data, enabling developers to quickly identify and address performance bottlenecks. Zipkin supports multiple data storage options, such as in-memory, MySQL, Cassandra, and Elasticsearch, and can be integrated with various programming languages and frameworks.

- **OpenTelemetry**: OpenTelemetry is a unified observability framework that provides APIs, libraries, agents, and collector services for capturing distributed traces and metrics from applications. It aims to standardize the collection and analysis of telemetry data across different platforms and tools. OpenTelemetry supports a wide range of programming languages and integrates seamlessly with popular observability tools such as Prometheus and Jaeger. It offers advanced features such as context propagation, distributed tracing, and metric collection, making it a versatile choice for modern cloud-native applications.

Figure 8.4 shows an example of how Newrelic's tracing dashboard allows us to look at the API calls in a distributed system. It shows the entire trace duration, as well as sections of it, such as the backend duration, root span duration, and more. Within **Backend duration**, it shows different process entry and exit points, which is useful in debugging performance issues with different processes:

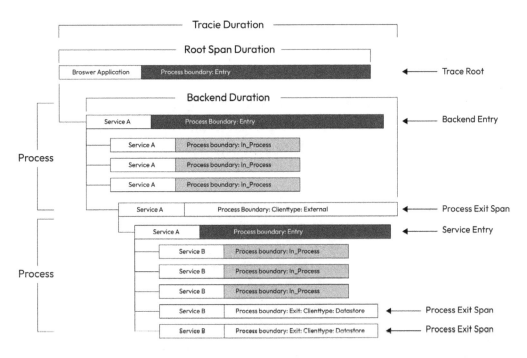

Figure 8.4: Newrelic tracing dashboard example

Best practices for implementing tracing

Implementing tracing in a distributed system can be challenging, but the following best practices can guide you through the process:

- **Instrument your code**: To collect traces, you need to instrument your code with the tracing library. This usually involves creating spans around operations you want to track.

- **Propagate context**: For a trace to span across multiple services, you need to propagate the trace context from one service to another. This is usually done by injecting the trace context into the headers of the HTTP request.

- **Use appropriate span names**: Span names should be descriptive and consistent so you can easily understand what operation they represent.

- **Annotate your spans**: Annotating spans with additional metadata can provide more context about the operation, making it easier to understand and troubleshoot issues.

Best practices

Throughout this chapter, we have explored key components of system design, including REST and gRPC APIs, API security, logging, metrics, alerting, and tracing in a distributed system. Each of these components plays a crucial role in building robust, scalable, and efficient systems.

As we conclude, let's summarize some best practices that can guide you in leveraging these components effectively:

- **Choose the right API design**: REST and gRPC each have their strengths and suitable use cases. Choose the one that best fits your system's requirements in terms of performance, ease of use, and compatibility.

- **Prioritize API security**: Implement robust authentication and authorization mechanisms, ensure secure communication, and use rate limiting to protect your APIs.

- **Implement centralized logging**: Collect logs from all your services and store them in a central location. This will make it easier to search and analyze logs, providing valuable insights into your system's behavior.

- **Monitor key metrics**: Identify key system, application, and business metrics, and monitor them regularly. This will help you understand your system's performance and make informed decisions.

- **Set up effective alerts**: Design alerts that are sensitive, specific, and provide enough context to diagnose issues. Regularly test your alerts to ensure they are working as expected.

- **Use distributed tracing**: Implement distributed tracing to track a request as it travels through various services. This will help you diagnose performance issues, understand the flow of requests, and optimize your system's performance.

Remember, the goal is not just to design a system that works, but to design a system that works well under a variety of conditions and can scale as your needs grow. By understanding and effectively leveraging these components, you will be well-equipped to design and build robust, scalable, and efficient systems.

Summary

In this chapter, we delved into the critical components of system design, covering APIs, security, logging, metrics, alerting, and tracing in distributed systems. We explored the nuances of REST and gRPC APIs, highlighting their distinct advantages and ideal use cases. The chapter emphasized the paramount importance of API security, outlining strategies for authentication, authorization, secure communication, and rate limiting.

We also examined the pivotal roles of logging and metrics in monitoring and debugging distributed systems, introducing centralized logging and the significance of key metrics. Alerting was discussed as a proactive measure for managing system incidents and ensuring timely response to potential issues. Furthermore, we explored tracing as a technique for tracking requests across services, providing invaluable insights into system performance and behavior.

By the end of this chapter, readers should gain a good understanding of the skills needed in designing and implementing these essential system components. These skills are crucial for building robust, scalable, and efficient systems, enabling readers to effectively monitor, troubleshoot, and optimize their systems in real-world scenarios. The lessons learned in this chapter lay the foundation for the subsequent chapters, where we will continue to build upon these concepts and explore advanced topics in system design.

In the next chapter, we will leverage all the basic components of system design that we have learned in the previous chapters to design real-world systems.

9

System Design – URL Shortener

In an era where brevity is valued and attention spans are fleeting, the quest for efficiency in digital communication has never been more crucial. **URL shorteners** have emerged as a vital tool in this landscape, offering a convenient solution to the age-old problem of unwieldy web addresses. But what goes into designing a URL shortener? Is it merely about creating a shorter link, or does it involve a more intricate interplay of technology, user experience, and ethical considerations?

The design of a URL shortener may seem straightforward at first glance, but beneath its seemingly simple facade lies a lot of complexities waiting to be unraveled. From choosing the right algorithms to ensure link uniqueness and stability, to crafting user interfaces that are intuitive and accessible, designing a URL shortener demands a multifaceted approach.

This chapter aims to delve deep into the intricacies of designing URL shorteners, understanding the functional and non-functional requirements, and exploring the challenges and considerations that developers and designers face in creating these indispensable tools. We'll examine the technical foundations that underpin URL shorteners, from URL parsing and hashing algorithms to database management and error handling.

We will cover the following topics in this chapter:

- Real-world use cases
- Functional requirements
- Non-functional requirements
- APIs
- Estimates and calculations
- System design
- Requirements verification

Join us as we embark on this journey of designing URL shorteners.

Real-world use cases

Before we start designing the URL shortener system, let's take a look at some of the real-world use cases where URL shorteners play a crucial role:

- **Social media sharing**

  - **Limited character count**: There may be limits to the number of characters in a post, which makes it challenging to share longer URLs

  - **Trackable links**: Brands and influencers use shortened URLs to track click-through rates and engagement metrics

- **Affiliate marketing**

  - **Track clicks**: Affiliate marketers use shortened URLs to track the performance of their campaigns by monitoring click-through rates

  - **Cleaner presentation**: Shortened URLs make promotional content look cleaner and more professional

- **Email marketing**

  - **Improved user experience**: Long URLs can clutter email content, making it less appealing and harder to read

  - **Track engagement**: Shortened URLs help marketers track which links recipients are clicking on and gauge the effectiveness of email campaigns

- **QR codes**

  - **Reduced complexity**: QR codes containing shorter URLs are simpler and quicker to scan

  - **Trackable**: Businesses can track the number of scans to measure the success of QR code-based campaigns

- **Print media and offline marketing**

  - **Memorable links**: Short URLs are easier for consumers to remember and type

  - **Track engagement**: Similar to other marketing channels, shortened URLs help measure the effectiveness of print media campaigns

- **Internal links**

  - **Ease of sharing**: Employees can share internal resources using shortened URLs, making collaboration more efficient

- **Track usage**: Organizations can monitor the usage of internal resources and identify popular content

- **Mobile applications**

    - **Improved user experience**: Shortened URLs can be more easily displayed within mobile apps without affecting the layout

    - **Dynamic linking**: Mobile apps use short URLs to deep-link to specific content within the app or direct users to the app store

- **Customization and branding**

    - **Branded links**: Companies can use custom domains to create branded short links, enhancing brand visibility and trust

    - **Consistent branding**: Custom short URLs maintain brand consistency across different marketing channels

By providing a more accessible, trackable, and user-friendly way to share and manage links, URL shorteners have become indispensable tools for marketers, businesses, and everyday internet users alike.

Now that we have understood the importance of the URL shortener service, let's dive into the *how* part of designing the system. Let's first brainstorm and document the functional and non-functional requirements.

Functional requirements

When about the URL shortener, we can think about a lot of functional requirements. It's important to identify a few core requirements among these. So, let's list out the core requirements first:

- Given a long URL, return a short URL

- Given the short URL, return the original long URL

- Short URLs should be unique and the length of the URL should have some cap

The following functional requirements can be considered extended or good to have requirements:

- Validate incoming long URL

- Support custom URLs

- Expiry of the short URL: 6 months based on the last usage

- Graceful handling

- The creator should be able to make changes to the long URL

- Deleting the URL mapping

- Support analytics and monitoring

- User account management

OK, this is a good prioritized collection of functional requirements. We need to go into the non-functional requirements.

Non-functional requirements

Having talked about the functional requirements, let's think about some non-functional requirements we should consider assuming we need to build this for a high scale:

- **Availability**: The system should be highly available

- **Scalability**: The system should be highly scalable (100 M users) and be able to tolerate request spikes

- **Latency**: The read and write requests should have very low latency

- **Consistency**:

  - Two users trying to access the long URL given a short URL should return the same long URL

  - Two users trying to create a short URL for the same long URL should get the same URL

  - One user created a short URL for a long URL and trying to create a short URL again for the same long URL should get the same short URL

- **Durability**: Once a short URL is created, we need to ensure that the data is never lost

- **Reliability**: The system should behave correctly and deliver the functional requirements even in case of failures, request spikes, and other outages

Now that we have a good understanding of the functional and non-functional requirements, let's document the APIs needed for this system to function.

Client APIs needed

The following APIs would be needed to support the core functional requirements of shortened URLs:

- `POST/shorturl`: Creates a new short URL given a long URL

  - `Request body`: The long URL

  - `Response`: The short URL

  - `getShortUrl(user_id, longurl)`

- `GET/shorturls/{shorturl}`: Returns the long URL corresponding to the short URL

  - Response: The long URL

  - `String getLongUrl(shortUrl)`

Next, let's try to better understand the scale of the problem by doing some back-of-the-envelope high-level calculations.

Estimates and calculations

For calculations, let's start by asking some questions:

- What characters can we use for short URLs? Our options are as follows:

  - a-z, A-Z, 0-9. So, it's 62 characters.

  - If you want to use more characters, -, ., _, and ~, so, there's a total of 66 characters.

  - Let's stick to just 62 characters.

- What's the scale we are operating at? Let's make some assumptions here:

  - We have 1 B users.

  - About 10% create short URLs (100 M users).

  - Each user creates 1 URL per day, so 100 M URLs per day.

  - We store these URLs for 10 years. So, `100 M * 1 * 365 * 10 = 365 B` URLs.

- How many characters would we have in the short URL? Now, to have 365 B unique URLs, how many characters do we need to create unique URLs? Let's see:

  - The number of URLs if we use 6 characters is `62^6 = 56 B` URLs, but our requirement is to have 365 B URLs. So, we need to use more characters.

  - If we use 7 characters, we have `62^7 = 3.5 T`.

- How much storage do we need? Let's see:

  - We have 365 B URLs. These are the columns we need to store for this system to work:

    - `short URL` (20 bytes)

    - `long URL` (1,000 bytes)

    - `created_at` (10 bytes)

- `updated_al` (10 bytes)
- `created_by` (20 bytes)

- So, overall, the number of bytes would be 1,060. Let's round it off to 1,500 bytes.
- So, the data we would store for each URL is 365 B* 1500 bytes ~= 500 TB.
- If we use a replication factor of 3, then we would need 1.5 PB of storage.

- What's the *read* and *write* RPS? Let's see:

  - **Write RPS**: 100 M requests per day = 100 M requests per day/(86,400 seconds/day) =~ 100 M/ 100,000 (approximating 86,400 to 100,000) requests per second = 1000 RPS
  - **Read RPS**: 100 x write RPS = 100 k RPS

The insights we've gained from the preceding calculations are that we definitely need multiple servers to support high read and write RPS. We also found out that we would need seven characters to be able to support the scale of URLs.

Let's now dive deeper into the design part of this chapter.

System design

In this section, we will explore solution options, bottlenecks, problems, mitigations, and high-level architecture diagrams. First, let's start with identifying what the core challenge here is.

Core challenge

The core challenge in this design problem is as follows: how do we generate unique URLs? There can be various solutions. Let's explore one at a time and see what are the pros and cons of each one and let's arrive at the most suitable one.

Option A – generate a random URL

Given a long URL, generate a random short URL and check whether exists in the **database** (**DB**). If it exists, then generate another random URL and keep doing it until we find that the entry doesn't exist. If it doesn't exist, then store this new entry in the DB as a key-value pair: {`short_url -> long_url`}.

There are a few problems with this approach:

- As you can see, there has to be at least one check done against the DB to find whether the entry already exists, but in some cases, it may multiply back and forth, generating a random URL and checking against the DB. This increases unpredictable delays and adds latency to the write requests.

- The other problem with this approach is that there may be collisions due to concurrency. Imagine a scenario where there are two requests trying to put the short URL at the same time. So, there are two requests, one for `long_url1` and another for `long_url2` and, coincidentally, the shortened URLs are the same for both the long URLs. Write for `long_url1` may win and `long_url2` may get corrupted. Here is the flow to understand this collision and corruption:

  - Both the short URLs generated for `long_url1` and `long_url2` are the same `short_url1`. So, we have `long_url1 → short_url1, long_url2 → short_url1`.

  - Now, the request for `long_url1` goes to check whether the `short_url1` in the DB is present or not. It is absent, but in the meantime, the request for `long_url2` also finds out that `short_url1` is absent, so it writes as an entry: {`short_url1 → long_url2`}.

  - Now, `long_url1 → short_url1` comes and overrides it and the entry becomes {`short_url1 → long_url2`}. So, `long_url2` is corrupted and lost.

Can we avoid this corruption?

It can be avoided by using the `putIfAbsent` kind of functionality, but not all DBs support this.

The other way to avoid this corruption is after writing, to do a `getLongUrl(short_url)` read call and check whether `long_url` is the one that the request was for. This can make sure it's not corrupted. It may take a few checks sometimes to be successful, and also, for every `put`, you are at least doing one `get`.

This is not the best solution to be deployed in production, so let's explore more.

Option B – MD5 hash

Given a long URL, generate its MD5 hash. The MD5 hash is 128 bits, and we need to have 7 bits with 62 base, which translates to 42 bits (base 2), since `62^7 ~= 2^42`. We can take 42 bits from the 128-bit MD5 hash and from that, we can create `short_url` with 7 base 62 characters.

One advantage of this is space saving; if the same URL is requested again, then you don't duplicate the short URL.

The problem with this approach is that since we are picking up fewer bits from the 128 bits, the probability of collision will be much higher. The first 42 characters of the MD5 hash of many URLs will be the same.

If we use more than 42 bits, then the collision probability is lower, but then that increases the number of characters in the short URL, which is not great. Also, unless we use all 128 bits for short URL generation, we can't completely avoid this collision.

Again, not a very good solution. Let's keep exploring other approaches.

Option C – counters approach

What if we generate a monotonically increasing counter incrementing the previous number by one?

So, a single host keeps generating this unique counter.

The problem with this approach is that this unique ID generator becomes a single point of failure. If this server dies, the whole system fails.

How can we mitigate this?

We can use multiple hosts, but making a counter unique across multiple servers is tough. What if we have the counter as host id (6 bits), time id (32 bits) + 4 bits (incrementing counter)?

There are a few problems with this approach:

- We will have 6 bits reserved for hosts, so we can only have 64 hosts
- The last 4 bits are for the incremental counter, so if the number of requests is more than 16 in a millisecond, then there would be collisions

This approach is also not ideal. Let's see whether the next solution is better.

Option D – distributed counters approach

The idea is to divide the 3.5 T sequence numbers into multiple ranges of 'k' size. Lets say, k is 1 million, thus, for 3.5T, we will have 3.5M such ranges, because each range is 1M itself.

So, we have 3.5 M such ranges, such as the following:

R1 [0-1M),

R2 [1M-2M),

...

R3 [9M-10M),

We can keep this in a highly available server (let's call it a Range server). We have multiple counter servers talking to this Range Server and asking for a range. A Range Server assigns a range and stores that in the following map. Let's assume there are 10 counter servers:

R1 [0-1M) → s1

R2 [1M-2M) → s2

...

R10 [9M-10M] → s10

The counter server increments the counter to generate a unique number across all the counter servers.

This unique number is now converted into a 7-character string (a-z, A-Z, 0-9) in the following procedure.

Our base 62-character set is abcdefghijklmnopqrstuvwxyzABCDEFGHIJKLMNOPQRS TUVWXYZ0123456789.

For example, say the unique number is 9234529445.

Then, we convert this base 10 number to a base 62 number:

9234529445 ÷ 62 = 148907051 remainder 23 → x

148907051 ÷ 62 = 2401748 remainder 39 → N

2401748 ÷ 62 = 38702 remainder 4 → e

38702 ÷ 62 = 624 remainder 14 → o

624 ÷ 62 = 10 remainder 4 → e

10 ÷ 62 = 0 remainder 10 → k

To summarize, the 9234529445 counter number in base 62 is {10, 4, 14, 4, 39, 23}, which, when mapped to the base 62 character set we chose, becomes keoeNx. So, the URL can be written as urlshortener.com/keoeNx.

This design looks good, but there are a few open questions:

- What happens if the range is exhausted by a counter server? This is what happens:

  When the range is exhausted by a server, it will ask for the next range from the Range Server.

- What if the counter server crashes? This is what happens:

  The range is discarded and we lose the entire 1 M range. This may be OK as we have 3.5 M ranges, so just one range getting discarded is not a big deal. Also, given that the counter server failing is not a frequent incident, this may be OK.

- How can we still prevent losing the entire range? Let's see:

  We can write to the Range Server periodically, for example, every Nth (let's say $N = 100$ or 1000) URL generation, so we can recover from the last entry + N. This way, we just discard only N counters at the maximum per server per failure incident.

- Isn't Range Server a single point of failure here? Let's see:

  Yes, it appears so, but since the Range Server doesn't have a lot of reads and writes, we can use a highly available, replicated, fault-tolerant service such as ZooKeeper.

This option looks much more robust than the previous options and doesn't have any major drawbacks.

Choice of database

If we look at the main data we are storing in this system, the design problem is the `short_url ->` `long_url` mapping. This is a key-value pair and an intuitive choice for storing this kind of mapping is Redis. Redis is a fast, highly scalable, durable, and fault-tolerant database.

Now that we have tackled the core challenge of this problem, let's summarize the solution in a high-level architecture diagram and flow.

High-level solution architecture

Figure 9.1 describes the high-level architecture and flow considering the *Option D* solution, which is the counter-range-based approach.

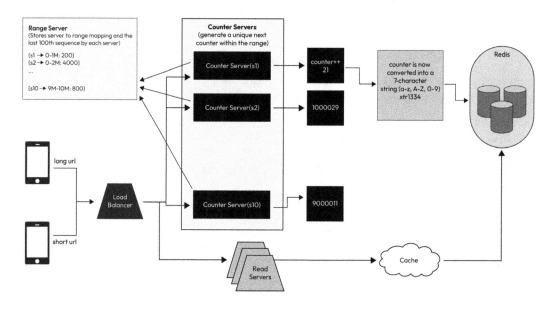

Figure 9.1: High-level architecture and flow

Figure 9.1 shows the following components: client machines, a load balancer, a range server, counter servers, a Redis database, read servers, and a cache. Let's take a look at the two high-level flows:

- **Write flow**

1. The client device calls the API to create a short URL given a long URL.
2. The load balancer routes this request to one of the counter servers, each having a unique range of counters that they got from the range server.
3. The assigned counter server generates the next counter by incrementing the last counter.

4. This new counter is then converted into a seven-character string, which is used to create the short URL.

5. This seven-character string and the long URL are stored in the Redis database as a `long_url -> short_url` mapping.

- **Read flow**

1. The client device calls the API to get the long URL for a particular short URL.

2. The load balancer routes the request to one of the read servers.

3. The assigned read server checks in the cache to see whether the particular short URL entry exists there.

   - If yes, then it returns the corresponding long URL from the cache.

   - If not, it fetches it from the Redis database.

   - The read server returns this long URL to the client device.

Now that we have the high-level architecture diagram and flows, let's verify whether all the functional and non-functional requirements are satisfied with the preceding final design.

Requirements verification

Let's quickly check whether all the functional and non-functional requirements are met with this design.

The following are the functional requirements:

- Given a long URL, return a short URL. This works based on our solution.

- Given the short URL, return the original URL. This requirement is also satisfied.

- Short URLs should be unique and the length of the URL should have some cap. The short URL is unique and is seven characters long.

The following are the non-functional requirements:

- **Availability**: The system should be highly available.

  - Given that the system is horizontally scalable and there is no single point of failure, there is always a server available to take the read or write requests.

- **Scalability**: The system should be highly scalable (100 M users) and be able to tolerate request spikes.

  - The system can be scaled by just adding more counter servers and read servers as they are behind a load balancer.

- **Latency**: The read and write requests should have very low latency:

  - **Write requests**: These should be fast as we don't do much calculation or database lookups. We can make it even faster by regenerating a few 100s of counters and hence the URLs and keeping them in memory.

  - **Read requests**: Redis has in-memory data and is very fast, but reads can be made even faster by putting a cache with LRU (Least Recently Used) as an eviction policy.

- **Consistency**: The system is highly consistent as there is only one entry per `short_url` stored in the database by design.

- **Durability**: Once a short URL is created, we need to ensure that the data is never lost. This is good since we are using Redis, which is durable.

- **Reliability**: The system should behave correctly and deliver the functional requirements even in case of failures, request spikes, and other outages. The system is reliable.

Summary

In this chapter, we designed the URL shortener system. We first looked at why it's a useful service to have and what some of the real-world use cases for such a service are. This helped us understand the importance of designing this system correctly. This was also useful to understand the scale at which such a system can be used.

We then started listing the requirements, both functional and non-functional. We classified some of the functional requirements as core and some as extended. Then, we went on to list the core APIs needed for this system to function. We also did some quick calculations to understand the scale, storage needs, and the number of characters needed in the shortened URL to serve the 3.5 T URLs.

After this, we started exploring various solution ideas to solve the core challenge, namely how to assign a unique shortened URL for a long URL. In part of the exploration, we identified various problems and tried to mitigate them by using some strategies. In the end, one of the solutions seemed to be most robust and applicable to our requirements of the system.

We put the whole solution together in an architectural diagram and depicted the flow. Finally, we went through the requirements and saw whether they were satisfied with our final solution.

In the next chapter, we will look into designing a proximity service used in services such as Uber and Yelp.

10

System Design – Proximity Service

From location-based marketing to smart transportation systems to finding a nearby restaurant or a business, **proximity services** leverage the power of proximity data to enhance user experiences, streamline operations, and unlock new opportunities across various industries.

In this chapter, we will understand the complexities of designing proximity services starting from stating the functional and non-functional requirements, to thinking about the core challenges, APIs, data storage, and serving the read and write requests. We'll also dive deeper into the technical concepts needed to break the problem down and form the foundations of the solution.

Understanding the principles behind proximity service design is essential for staying ahead in the rapidly evolving landscape of digital transformation. This chapter attempts to delve into the details of creating a large-scale system based on the proximity service.

We will cover the following topics in this chapter:

- Real-world use cases
- Functional requirements
- Non-functional requirements
- Client APIs needed
- Estimates and calculations
- System design
- Requirements verification

So, let's start the chapter by considering some real-world use cases for a proximity service.

Real-world use cases

Proximity services play a pivotal role in various popular applications that people use daily. Here are some real-world use cases of proximity services exemplified by well-known platforms:

- **Ride-sharing services (Uber, Lyft):**

  - Driver matching: These platforms use proximity services to match riders with nearby drivers efficiently. When a user requests a ride, the system identifies and notifies drivers in close proximity to the pick-up location, optimizing response times and reducing wait times for passengers.

  - Dynamic pricing: Proximity data helps adjust fares based on demand and driver availability in specific areas, ensuring a balance between supply and demand and providing incentives for drivers to operate in high-demand zones.

- **Local search and recommendation services (Yelp, Foursquare):**

  - Location-based recommendations: These platforms leverage proximity services to deliver personalized recommendations for restaurants, shops, attractions, and services based on a user's current location. Users receive suggestions for nearby places tailored to their preferences, ratings, and reviews.

  - Check-in and loyalty programs: Proximity-based check-in features enable users to notify friends about their location, share experiences, and participate in loyalty programs or promotions offered by local businesses.

- **Food delivery services (DoorDash, Grubhub):**

  - Optimized delivery routes: Proximity services help drivers optimize delivery routes by identifying the most efficient path to reach multiple destinations in a single trip, reducing delivery times and improving overall efficiency.

  - Real-time tracking: Users can keep track of the status of their orders in near-real time, getting updates on the food preparation, pick-up readiness, actual pickup, and estimated arrival time of their food delivery based on the proximity of the driver to their location.

- **Social networking and dating apps (Tinder, Bumble):**

  - Geolocation matching: These apps utilize proximity services to connect users with potential matches in their vicinity, allowing users to view profiles, initiate conversations, and arrange meetings with people nearby.

  - Event and venue discovery: Proximity-based features enable users to discover local events, gatherings, and venues, facilitating social interactions and networking opportunities based on shared interests and location.

- **Navigation and mapping services (Google Maps, Waze):**

  - Turn-by-turn directions: Proximity services provide accurate location data to deliver real-time, turn-by-turn navigation instructions, helping users navigate to their destinations efficiently and safely.

  - Traffic and incident alerts: These platforms use proximity data to monitor traffic conditions, identify congestion, accidents, or road closures in the vicinity, and provide alternative routes to avoid delays.

- **Fitness and health tracking apps (Strava, Fitbit):**

  - Location-based activity tracking: Proximity services enable these apps to track and analyze users' activities, such as running, cycling, or walking, providing insights into performance metrics, route mapping, and personalized fitness recommendations based on the proximity to popular routes or landmarks.

These examples showcase the diverse applications of proximity services across different industries, highlighting their role in enhancing user experiences, optimizing operations, and driving innovation in today's digital landscape. Whether it's connecting people, delivering services, or facilitating interactions based on location data, proximity services continue to redefine how we engage with the world around us.

Having explored real-world examples of proximity services, let's now delve into the process of designing such a system. First, we'll brainstorm and outline both the functional and non-functional requirements, keeping a simple use case in mind – finding nearby restaurants and placing an order.

Functional requirements

Here are the core requirements for this design problem:

- A user should be able to search nearby restaurants, given their location
- A user should be able to select a restaurant and place an order

Non-functional requirements

Let's put down the non-functional requirements for this design problem:

- **Availability**: The system should be highly available.
- **Scalability**: The system should be highly scalable (100M users) and be able to tolerate request spikes.
- **Latency**: Since restaurants' menus don't change often and restaurants don't open and close very frequently, there are not many write requests happening. Read latency should be within 200 ms.
- **Consistency**: Occasionally, when the restaurants are updated, it's fine to support eventual consistency.

- **Reliability**: The system should behave correctly and deliver the functional requirements even in the case of failures, request spikes, and other outages.

Client APIs needed

Now that we have thought about the functional and non-functional requirements, let's document the APIs needed for the functional requirements to be satisfied:

1. `GET /restaurants/search?lat=37.7749&long=122.4194&distance=5`

```
Response:
[
  {
    "restaurantId": "78566",
    "name": "ABC_Italian",
    "location": "123 Castro St, San Francisco, CA 94056",
    "cuisine": "Italian",
    "rating": 4.9,
    "distance": 2 miles
  },
  {
    "restaurantId": "45678",
    "name": "DEF_Mexican",
    "location": "123 Main St, San Francisco, CA 94056",
    "cuisine": "Mexican",
    "rating": 4.5,
    "distance": 3 miles
  }
  ...
]
```

2. `POST /orders/place`

```
Request:
{
  "userId": "12345",
  "restaurantId": "78566",
  "items": [
    {
      "itemId": "item1",
      "quantity": 4
    },
```

```
      {
         "itemId": "item2",
         "quantity": 3
      }
   ],
   "paymentMethod": "credit_card"
}


Response:
{
   "orderId": "6689092",
   "message": "Thanks for ordering with us!",
   "Bill Amount": $75.92
}
```

Now that we have thought through the APIs needed, let's try to estimate the scale of the problem by doing some back-of-the-envelope calculations.

Estimates and calculations

Let's start by asking and answering some questions:

- How many users do we have? 100M

- How many **Daily Active Users (DAU)** do we have? 10% of total users, making 10M/day.

- Queries per second to support for daily active users: People usually order food around lunch and dinner time, so let's assume that orders will be placed for 3 hours during lunch and 3 hours at dinner time. 10M DAU placing orders within 6 hours means that QPS = 10M/6*60*24 ~= 1200 QPS.

- What about traffic spikes? I think 5x is a good assumption: 5 * 1200 ~ 60000 QPS.

- How many people are browsing restaurants and doing searches? Assuming 10M users doing 5 search queries, then search QPS = ordering QPS *5 = 60000 QPS.

- How many restaurants are on the app? Let's assume 10M restaurants in the world and also let's assume 1M restaurants are online on our app. Restaurant data doesn't change often.

OK, now we have the high-level estimates done. These tell us that our system will have a high QPS with spikes, so we should design our system to be scalable and tolerate spikes in demand. Let's move on to the actual core part of designing the system.

System design

In this section, we will design the system. First, we will put together a high-level system design with different components and design the flows that satisfy the functional requirements. Then we will identify the single points of failure or any issues with our design and list out the core challenges that we need to address. This will help us refine the design and arrive at the final high-level system design architecture and flow.

Before we dive deep into the core challenges, let's put a high-level diagram first to put together the pieces and the flow.

High-level system design

The following diagram shows the initial high-level design with the important entities and the flow (*Figure 10.1*). The core entities we discuss here are **Customer**, **Restaurant**, and **Order**. The following high-level diagram just shows the basic components and simple databases to store the data. We will refine it later as we dig deeper.

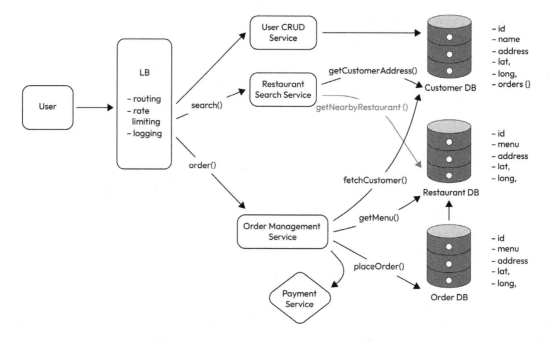

Figure 10.1: High-level initial design diagram

Core challenge

The challenge in this design problem is finding nearby restaurants given the users location (their latitude and longitude) and the given radius. We will refer to latitude as **lat** and longitude as **long** henceforth. So the user location is *(lat, long)*. Figure 10.1 depicts this with the `getNearbyRestaurants()` call from **Restaurant search service** to **Restaurant DB**. Let's explore various options to solve this.

Option A

We can store the restaurant details along with its lat and long in a relational DB and do a search using an SQL query with the user's lat and long values (`user_lat`, `user_long`), as follows:

SQL query

```
select restaurant_ids from restaurant_table where
    lat > user_lat - 5 AND lat < user_lat + 5 AND
    long > user_long - 5 AND long < user_long + 5
```

Problem: This query is very expensive and scans the entire table.

Can we build indexes?

Indexes are good in one dimension, allowing us to do a sort of binary search, but not in two dimensions. So in this case, even if we build two indexes, we will be able to use only one index at a time. This narrows down the search from the entire world to a search in the range of *long +/- 5 miles*, which is also huge and not very efficient.

So let's explore other options.

Option B

Using quadtree

A **quadtree** is a tree similar to a binary tree, except that it has 4 children.

- The world is divided into quadrants. Quadrants are formed as and when we see a quadrant having less than N number of restaurants (where N=500, let's say), as shown in *Figure 10.2*.
- Each of these quadrants is a node in a quadtree.
- The leaf node contains the list of restaurant IDs.
- The non-leaf nodes contain the min and max lat and long values.

The following diagram shows how the whole world map can be divided into recursively smaller rectangular quadrants.

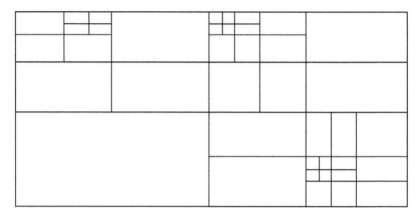

Figure 10.2: World map divided into quadrants recursively with threshold checks

Figure 10.3 shows a quadtree. As we know, it's a tree with four child nodes. The **Root** node here represents the whole world map, and the four children of the root are the four big quadrants when the world map rectangle is divided into four equal rectangles. We represent each rectangle as a node in *Figure 10.2*. Each node contains the min and max lat and long values. The leaf nodes of this tree contain the list of given restaurants inside the quadrant in the world map.

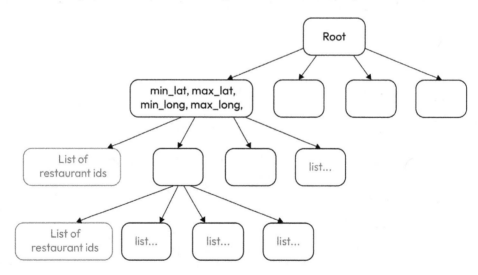

Figure 10.3: quadtree structure

Let's use the quadtree in *Figure 10.3* to see how we find the nearby restaurants given the user's lat and long and a radius =D.

Here is the high-level algorithm:

1. Start from Root.

2. Check whether the (lat, long) value is within the node's min and max limits.

3. If yes, do the DFS (Depth First Search) for each of the four children and repeat *steps 1* and *2* until you land on a leaf node.

4. Now we are on a leaf node:

    *   Get the list of restaurants

    *   Apply the Cartesian or driving distance filter of distance D from the user's lat and long

    *   Return the final list of restaurants

It is possible that we find out that the list is not sufficient and we need to expand the radius of the search. So, we can just go one level up to the parent node of this current node and then visit each of the other three children of the node's parent. The following are the pros and cons of this:

*   Pros: Fast as it's in memory data structure

*   Cons: If there is a need for frequent changes, it would be difficult to change the tree all the time

This may be an acceptable tradeoff since restaurants don't open and close very often. Let's discuss another popular approach in this space.

Option C

Geohashing converts latitude and longitude into a single, shorter string. Lets explore using this solution option.

*   The world is divided into quadrants as we discussed earlier, but now, we label each quadrant as shown in *Figure 10.4*

*   The top left quadrant is 00, the top-right one is 01, the bottom left is 10, and the bottom right is 11

*   We recursively divide all the quadrants and label them recursively

So the first quadrant 00 is further divided into 4 quadrants called 00, 01, 10, and 11, as discussed previously, and hence the label for these smaller quadrants will be 0000, 0001, 0010, and 0011

The following *Figure 10.4* shows how the world map is divided recursively into smaller rectangular areas. Contrary to the previous quadtree solution where the world map was divided conditionally if the number of restaurants was above a threshold, with geohashing, we don't consider any threshold.

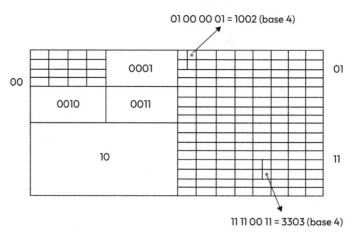

Figure 10.4: Entire world map divided into quadrants recursively with no thresholds

So imagine the whole world map is 25,000 miles x 12,500 miles. If we divide this into 16 recursive levels, we will have rectangles of .38 miles x 0.19 miles. Each of these rectangles will have a different 16-digit label in base 4. We can store it in base 4 or base 32, which is how it's usually stored.

We can use the numerals 0-9 and all lower case alphabet letters except (a, i, l and o) for the base 32 representation. We skip these 4 letters since we need only 22 and these are good candidates as they are a bit similar looking (i and l look a bit similar and so as a and o):

```
[0, 1, 2, 3, 4, 5, 6, 7, 8, 9, b, c, d, e, f, g, h, j, k, m, n, p,
q, r, s, t, u, v, w, x, y, z]
```

For example, say we have a quadrant with the location 0231 0101 0131 0131 (in base 4), then we can convert this into base 10 first, which comes out as 121610608. Next, let's convert this into a base-32 value as follows:

```
121610608 / 32 = 3794087 remainder 24 (s)

3794087 / 32 = 118565 remainder 7 (6)

118565 / 32 = 3705 remainder 25 (t)

3705 / 32 = 115 remainder 25 (t)

115 / 32 = 3 remainder 19 (m)

3 / 32 = 0 remainder 3 (2)

//
```

So the 16-digit base-4 number `0231 0101 0131 0131 = 2mtt6s`.

> **Coordinates of San Francisco city is**
>
> (37.7749° N, 122.4194° W) Converting it into base 32 geohash is
>
> Looking at this site: https://www.dcode.fr/geohash-coordinates the coordinates of San Francisco city with precision up to 4 digits is 9q8yyk8yt

All of the restaurants in this quadrant will thus have the prefix `9q8yyk8yt`. So we can store the location and list of restaurant IDs in a relational table similar to the following.

geohash	restaurant_ids
9q8yyk	{3, 7, 97, 89, 234...}
9q8yym	{1, 4, 73, 91, 212...}
9q8yym	{9, 13, 92, 893, 422...}
...	...

Table 10.1: database table storing the geohash and corresponding list of restaurant IDs in that geohash location.

Let's say the user's lat long was 37.7749° N, 122.4194° W. The geohash of this is `9q8yyk`. When this user searches for restaurants in their local area, the query is as follows:

```
select restaurant_ids where geohash like "9a8yy%"
```

This will return the geohashes of all the restaurants with the prefix 9q8yy, which are near to the user's location as represented by the geohash (`9q8yyk`).

Making the right choice for the solution

A simple relational DB solution is not appropriate for this problem, but both the quadtree and geohashing solutions seem to work fine. Which one is better?

Let's consider one scenario – an area that doesn't have many restaurants.

Geohashing will continue to divide into additional levels, which may result in unnecessary subdivisions with few or no restaurants, leading to wasted space and increased time to retrieve matching restaurants. In contrast, quadtree avoids further subdivision if the number of restaurants is less than N (e.g., 500), making it more efficient. Another significant advantage of quadtree over Geohashing is its in-memory structure, which allows for very fast access times.

However, a key drawback of quadtree is the complexity of frequent updates, as it requires serialization and data persistence to handle crashes and rebuild the in-memory tree. Given that restaurant locations do not change frequently in our use case, this tradeoff is acceptable.

This brings us to the question of how we populate the restaurant quadtree. We can populate it whenever a restaurant closes or a new one opens up. Since this doesn't happen very frequently, this sync update works fine.

OK, at this point, we have addressed some of the core challenges we had identified and now we are ready to put the whole design together.

Final high-level solution architecture

The core problem was to get the nearby restaurant data in a scalable and performant way. As discussed in the previous section, we will use the quadtree for our solution. *Figure 10.5* shows the final high-level system diagram.

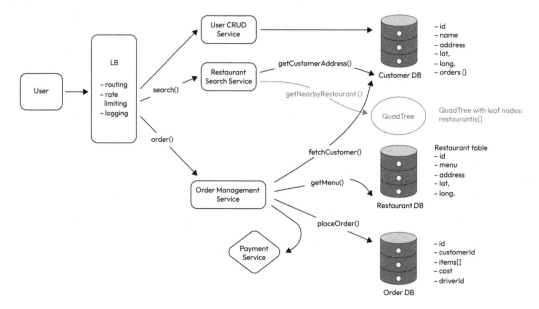

Figure 10.5: Final high-level system diagram

Let's take a look at the high-level flows.

Write flow:

1. The **User** device calls the API to do **Create, Read, Update, and Delete (CRUD)** operations. The load balancer redirects it to the **User CRUD** service, which makes the changes to the Customer DB.

2. The **User** device calls an API to place the order. The load balancer routes this request to the order management service, which in turn fetches the required customer data, gets the menu from the restaurant database, and then helps the customer select the items and place the order. The order management service also talks to the payment service to take care of the payment and when done, it finally writes the order entry in the order database.

Read flow:

1. The **User** device calls the API to search for nearby restaurants.

2. The load balancer routes the request to the restaurant search service, which talks to the quadtree and goes from the root node to the leaf node based on the customer address and their lat and long. It then returns the list of nearest k restaurants, which in turn is finally returned to the user.

This concludes our system design work. Now, let's validate our functional and non-functional requirements against this design in the next section.

Requirements verification

Let's quickly check whether all of our functional and non-functional requirements are met with this design.

The following are the two functional requirements we had:

- A user should be able to search nearby restaurants, given their location.

- A user should be able to select a restaurant and place an order.

Both these are satisfied by the current design.

The following are the non-functional requirements:

- **Availability**: The system should be highly available.

  The system is highly available since we will have all these services have multiple instances running, so even if one or a few go down, the system doesn't go down and is still serving the traffic.

- **Scalability**: The system should be highly scalable (100M users) and be able to tolerate request spikes.

  This is a horizontally scalable system with no bottlenecks.

- **Latency**: Read latency should be within 200 ms.

  Latency will be < 200 ms since we will use quadtree for the search use case.

- **Consistency**: Occasionally, when the restaurants are updated, it's fine to support eventual consistency.

  Eventual consistency is fine here.

- **Reliability**: The system should behave correctly and deliver the functional requirements even in the case of failures, request spikes, and other outages.

  We can see that the system is durable, reliable, and fault-tolerant. The non-functional requirements are therefore also satisfied.

Summary

Proximity services leverage proximity data to enhance user experiences, streamline operations, and unlock new opportunities across various industries. This chapter delves into the complexities of designing proximity services, starting from defining functional and non-functional requirements to addressing core challenges and technical concepts. Real-world use cases highlight the diverse applications of proximity services, ranging from ride-sharing and local search to food delivery and social networking apps.

The chapter outlines the process of designing a proximity service, focusing on functional requirements such as searching nearby restaurants and placing orders, and non-functional requirements including availability, scalability, latency, consistency, and reliability. APIs for restaurant search and order placement are provided along with estimates and calculations for handling traffic spikes and user interactions.

We then moved on to solving the core challenge in this use case, which was how to find nearby restaurants. Two approaches for finding nearby restaurants were discussed: quadtree and Geohashing. While both solutions offer advantages, quadtree was preferred for its in-memory nature and efficiency in handling frequent updates. The final high-level solution architecture incorporated these concepts to meet all functional and non-functional requirements, ensuring high availability, scalability, low latency, eventual consistency, and reliability.

The chapter concluded by verifying that the proposed solution met all of our requirements, making it a robust and effective design for proximity services.

In the next chapter, we will learn about the system design of the Twitter app.

11

Designing a Service Like Twitter

In today's digital landscape, social media platforms have revolutionized communication and information sharing. Twitter, currently know as X, is a microblogging service allowing users to share short messages called tweets, has emerged as a global phenomenon with millions of active users and billions of tweets generated daily. Designing a service like Twitter presents unique challenges in scalability, reliability, and user experience.

This chapter explores the system design of a Twitter-like service, applying our learnings from basic system design blocks to create a scalable and efficient platform. We will examine the core features, non-functional requirements, data models, and scale calculations that form the foundation of the system. Based on these, we'll propose a high-level design architecture leveraging various components, such as load balancers, API gateways, caches, databases, and storage systems. We'll also delve into the low-level design of key services, emphasizing scalability, reliability, and performance throughout.

We will cover the following topics in this chapter:

- Functional requirements
- Non-functional requirements
- Data model
- Scale calculations
- Designing Tweet Service
- Designing User Service
- Low-level design –Timeline Service
- Designing Search Service
- Additional considerations

- Low-level design of key components (Tweet Service, User Service, Timeline Service, and Search Service)

- Scalability techniques (caching, sharding, and asynchronous processing)

By the end of this chapter, you will have a comprehensive understanding of the principles and practices involved in designing a scalable and robust social media platform like Twitter.

Let's explore the key functional requirements.

Functional requirements

Here are the functional requirements for our system:

- User registration and authentication:

  - Users should be able to create new accounts by providing the necessary information, such as username, email, and password

  - The system should securely store user credentials and authenticate users upon login

  - User sessions should be managed efficiently to allow seamless access to the service

- Tweeting:

  - Users should be able to post tweets, which are short messages limited to a specific character count (e.g., 280 characters)

  - Tweets can contain text, hashtags, mentions of other users, and media attachments such as images or videos

  - The system should enforce the character limit and handle the storage and retrieval of tweets efficiently

- Follow/unfollow:

  - Users should be able to follow other users to receive their tweets in their timeline

  - The system should maintain a follow graph that represents the relationships between users

  - Users should also have the ability to unfollow other users, removing their tweets from the timeline

- Timeline:

  - The timeline is a crucial feature that displays a chronological feed of tweets from the users a person follows

  - The system should efficiently generate and serve personalized timelines for each user, considering factors such as tweet timestamps and user preferences

- The timeline should support real-time updates, ensuring that users see the latest tweets as they are posted

- Searching:

  - Users should be able to search for tweets and other users based on keywords, hashtags, or usernames

  - The search functionality should provide relevant and accurate results, considering factors such as relevance, popularity, and recency

  - The system should efficiently index and store tweet and user data to enable fast and scalable searching

- Retweeting and liking:

  - Users should have the ability to retweet, which means sharing another user's tweet with their own followers

  - The system should handle retweets efficiently, maintaining the original tweet's metadata and attribution

  - Users should also be able to like tweets, indicating their appreciation or agreement with the content

- Direct messaging:

  - The service should support direct messaging functionality, allowing users to privately communicate with each other

  - Users should be able to send and receive direct messages, which are separate from the public tweet stream

  - The system should ensure the privacy and security of direct messages, implementing proper access controls and encryption

These functional requirements provide a comprehensive overview of the core features that a Twitter-like service should offer. By fulfilling these requirements, the system will enable users to engage in microblogging, connect with others, and consume relevant content in real-time.

In the next section, we will explore the non-functional requirements that ensure the service remains scalable, reliable, and performant while meeting the functional requirements discussed previously.

Non-functional requirements

While functional requirements define what the system should do, non-functional requirements specify how the system should perform and behave. These requirements are critical to ensuring that the Twitter-like service remains scalable, available, and reliable under various conditions. Let's discuss the key non-functional requirements:

- Scalability:

  - The system should be designed to handle a large number of users and tweets, accommodating growth and peak traffic loads

  - Horizontal scalability should be achieved by adding more servers and distributing the load across them

  - The architecture should allow for easy scaling of individual components, such as Tweet Service or Timeline Service, independently

- Availability:

  - The service should be highly available, ensuring minimal downtime and quick recovery from failures

  - Redundancy should be implemented at various levels, including server redundancy, database replication, and geo-redundancy

  - The system should be designed to handle server failures, network outages, and data center disasters without significant impact on user experience

- Reliability:

  - The system should be reliable, ensuring data integrity and consistency across all components

  - Mechanisms should be in place to prevent data loss, such as regular backups and data replication

  - Consistency models should be chosen carefully to balance data accuracy and performance, considering factors such as eventual consistency or strong consistency

- Latency:

  - The service should provide real-time updates and fast response times to ensure a smooth user experience

  - Latency should be minimized for critical operations, such as posting tweets, viewing timelines, and receiving notifications

  - Techniques such as caching, **content delivery networks** (CDNs), and efficient data retrieval should be employed to reduce latency

By addressing these non-functional requirements, the Twitter-like service can ensure a reliable, scalable, and performant user experience. It is essential to consider these requirements throughout the design process and make architectural decisions that align with these goals.

Let us now learn about the data model that forms the foundation of the Twitter-like service, defining the entities and relationships necessary to support the functional requirements.

Data model

The data model is a crucial component of the Twitter-like service, as it defines the structure and relationships of the data entities involved. A well-designed data model ensures efficient storage, retrieval, and manipulation of data while supporting the functional requirements of the system. Let's dive into the key entities and their relationships. *Figure 11.1* captures the UML-style class diagram for the different data models and how they interact with each other.

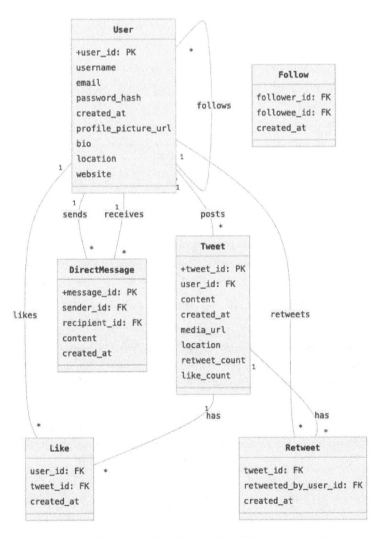

Figure 11.1: UML-style class diagram for different data models

Figure 11.1's UML-style class diagram provides a comprehensive visual representation of the data model for our Twitter-like service. It illustrates the key entities (User, Tweet, Follow, Like, Retweet, and DirectMessage) along with their attributes and the relationships between them. The diagram clearly depicts the one-to-many and many-to-many relationships, such as users posting multiple tweets or following multiple users, offering a clear overview of the system's data structure and relationships at a glance.

At a high level, in our system, we need to represent users, their tweets, and their relationships with other users. Our data models will be designed to speed up the common queries and will support the uncommon ones. Let's summarize the different entities and relationships between them:

- The User entity represents the users of the Twitter-like service, storing their profile information such as username, email, password hash, profile picture, bio, location, and website. Each user is uniquely identified by their user_id.

- The Tweet entity represents the tweets posted by users. It contains the tweet content, user ID of the author, creation timestamp, media URL (if applicable), location, and counts for retweets and likes. tweet_id serves as the primary key.

- The Follow entity represents the follow relationships between users. It contains follower_id and followee_id, indicating which user is following whom. The created_at timestamp captures when the follow relationship was established.

- The Like entity represents the likes on tweets. It contains the user ID of the user who liked the tweet and the tweet ID of the liked tweet. The created_at timestamp records when the like occurred.

- The Retweet entity represents the retweets of tweets. It contains the tweet ID of the original tweet and retweeted_by_user_id, which is the ID of the user who retweeted the tweet. The created_at timestamp indicates when the retweet happened.

- The Direct Message entity represents private conversations between users. It contains message_id, sender_id, recipient_id, content, and the created_at timestamp.

By designing the data model with these entities and relationships, the Twitter-like service can efficiently store and retrieve data related to users, tweets, follows, likes, retweets, and direct messages. The data model supports the functional requirements and enables the system to handle the complex interactions between users and their activities.

To build a usable system, it is important to understand the scale at which it will operate. Now, we will perform scale calculations to estimate the storage, bandwidth, and processing requirements of the Twitter-like service based on the anticipated user base and usage patterns.

Scale calculations

To design a scalable Twitter-like service, it is essential to estimate the storage, bandwidth, and processing requirements based on the expected user base and usage patterns. These calculations help in making informed decisions about the infrastructure and resources needed to support the service. Let's perform some scale calculations:

- Assumptions:

  - Total number of users: 100 million

  - Daily active users: 20 million

  - Average number of tweets per user per day: 5

  - Average tweet size: 200 bytes

  - Average media size per tweet: 1 MB

  - Percentage of tweets with media: 20%

  - Retention period for tweets: Five years

- Storage requirements:

  - Tweet storage:

    - Daily tweet storage: `20 million users × 5 tweets/user/day × 200 bytes/tweet = 20 GB/day`

    - Yearly tweet storage: `20 GB/day × 365 days = 7.3 TB/year`

    - Total tweet storage for five years: `7.3 TB/year × 5 years = 36.5 TB`

- Media storage:

  - Daily media storage: `20 million users × 5 tweets/user/day × 20% media tweets × 1 MB/media = 200 TB/day`

  - Yearly media storage: `200 TB/day × 365 days = 73 PB/year`

  - Total media storage for five years: 73 PB/year × 5 years = 365 PB

- User storage – assuming 1 MB of storage per user (profile picture, bio, etc.):

  - Total user storage: `100 million users × 1 MB/user = 100 TB`

  - Total storage: `36.5 TB (tweets) + 365 PB (media) + 100 TB (users)` `≈ 365 PB`

- Bandwidth considerations:

  Daily bandwidth for tweet delivery: Assuming an average of 100 followers per user:

  - Daily tweet deliveries: `20 million users × 5 tweets/user/day × 100 followers/user = 10 billion tweet deliveries/day`

  - Daily bandwidth: `10 billion tweet deliveries/day × 200 bytes/tweet = 2 TB/day`

- Daily bandwidth for media delivery:

  - Daily media deliveries: `20 million users × 5 tweets/user/day × 20% media tweets × 100 followers/user = 2 billion media deliveries/day`

  - Daily bandwidth: `2 billion media deliveries/day × 1 MB/media = 2 PB/day`

  - Total daily bandwidth: `2 TB (tweets) + 2 PB (media) ≈ 2 PB/day`

- Processing requirements:

  - Peak tweets per second: `20 million users × 5 tweets/user/day ÷ 86400 seconds/day ≈ 1,200 tweets/second`

  - Peak media uploads per second: `1,200 tweets/second × 20% media tweets ≈ 240 media uploads/second`

  - Fanout requests for timeline generation: `100 followers/user × 1200 tweets/second ≈ 120,000 requests/second`

- Cache sizing:

  Assuming 80% of the daily tweet views can be served from the cache

  - Daily tweet views: `20 million users × 100 timeline views/user/day = 2 billion tweet views/day`

  - Cache size: `2 billion tweet views/day × 80% cache hit rate × 200 bytes/tweet ≈ 320 GB`

These calculations provide a rough estimate of the storage, bandwidth, and processing requirements for the Twitter-like service. It's important to note that these numbers can vary based on the actual usage patterns and growth of the user base. The infrastructure should be designed to handle peak loads and should be easily scalable to accommodate future growth.

Before we dive into developing services and features, a good step is to get a high-level design in order so we can understand different modules and how they interact. The subsequent section will focus on designing various building blocks to create a scalable and efficient system.

Exploring high-level design

Now that we have a clear understanding of the functional and non-functional requirements, as well as the scale calculations, let's dive into the high-level design of the Twitter-like service. The goal is to create an architecture that is scalable, reliable, and efficient in handling the massive volume of tweets, users, and interactions. *Figure 11.2* shows the high-level design of the Twitter-like system, which includes a load balancer, API gateway, microservices such as User Service and Tweet Service, database tables, the caching layer, Kafka, and an object store. We will discuss these components at a high level in this section and in the next section do a deep dive into some of them.

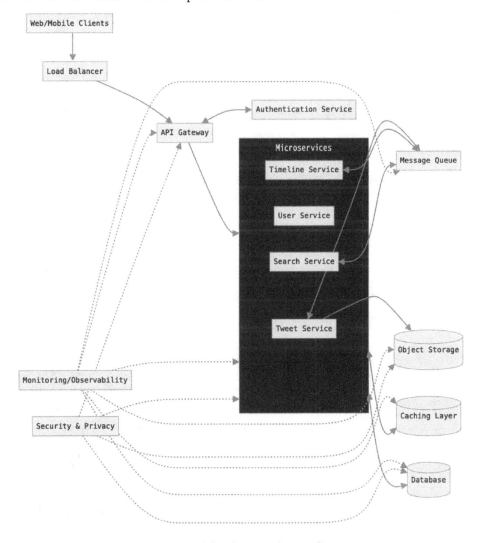

Figure 11.2: High-level system design of Twitter

Figure 11.2 shows different modules and microservices that need to be developed to build a system like Twitter. These are listed as follows, along with a brief explanation for each:

- **Client-server architecture:** The Twitter-like service will follow a client-server architecture, where clients (such as web browsers or mobile apps) communicate with the server-side components through APIs. The server-side components will handle the core functionality, data storage, and processing.

- **Load balancer:** To distribute the incoming traffic evenly across multiple servers, a load balancer will be placed in front of the server-side components. The load balancer will ensure that requests are efficiently routed to the appropriate servers based on factors such as server load, request type, and geographic location.

API gateway: An API gateway will act as the entry point for all client requests. It will handle request routing, authentication, rate limiting, and request/response transformation. The API gateway will expose well-defined APIs for various functionalities, such as tweeting, following, liking, and searching.

Microservices architecture

To promote modularity, scalability, and maintainability, the server-side components will be designed as microservices. Each microservice will be responsible for a specific domain or functionality. The main microservices in the Twitter-like service will be the following:

- Tweet Service:

  - Handles tweet creation, retrieval, and deletion

  - Stores tweet data in the database and media files in the object storage

  - Publishes new tweets to the message queue for processing by other services

- User Service:

  - Manages user registration, authentication, and profile information

  - Stores user data in the database

  - Handles the follow/unfollow functionality and maintains the follower-followee relationships

- Timeline Service:

  - Generates and serves user timelines by aggregating tweets from followed users

  - Consumes new tweets from the message queue and updates user timelines accordingly

  - Stores timeline data in the cache for fast retrieval

- Search Service:

  - Enables searching for tweets and users based on keywords, hashtags, and other criteria

  - Indexes tweet and user data for efficient searching

  - Provides search results ranked by relevance and popularity

In order to support performance, scale, and reliability, we need to develop a bunch of common software modules and themes that can be reused across different microservices. These cover caching, data models, the use of message queues, handling updates, monitoring and logging, and security and privacy, which we cover next:

- **Caching**: To improve performance and reduce the load on the backend services, a distributed caching layer (e.g., Redis) will be employed. The caching layer will store frequently accessed data, such as user profiles, popular tweets, and timeline data. Caching will help in serving data quickly and reducing the number of requests to the database.

- **Database**: The Twitter-like service will use a distributed database (e.g., Apache Cassandra or Amazon DynamoDB) to store structured data, such as tweets, users, follows, likes, and retweets. The database will be designed to handle high write throughput and provide low-latency reads. Techniques such as sharding and replication will be used to distribute the data across multiple nodes and ensure availability.

- **Object storage**: Media files, such as images and videos, will be stored in an object storage system (e.g., Amazon S3). Object storage provides scalable and durable storage for large files, allowing for efficient retrieval and delivery to users.

- **Message queue**: A message queue (e.g., Apache Kafka) will be used for asynchronous communication between microservices. When a new tweet is created, it will be published to the message queue. Timeline Service and Search Service will consume these messages and update their respective data stores accordingly. The message queue ensures loose coupling between services and enables scalable processing of tweets.

- **Real-time updates**: To provide real-time updates to users, WebSocket connections can be established between the clients and the server. When a new tweet is posted or an event occurs (e.g., a new follower), the server can push the updates to the relevant clients through the WebSocket connection, ensuring instant notification.

- **Monitoring and logging**: Comprehensive monitoring and logging mechanisms will be implemented to track the health and performance of the system. Metrics such as request latency, error rates, and resource utilization will be collected and visualized using tools such as Prometheus and Grafana. Centralized logging solutions (e.g., ELK stack) will be used to aggregate and analyze logs from all components.

- **Security and privacy**: Security and privacy measures will be implemented throughout the system. User authentication and authorization will be handled using secure protocols such as OAuth. Sensitive data, such as passwords, will be hashed and stored securely. Data encryption will be applied to protect user information both in transit and at rest. Rate limiting and throttling mechanisms will be put in place to prevent abuse and ensure fair usage of the service.

This high-level design provides an overview of the key components and their interactions in the Twitter-like service. It takes into account the scale requirements and employs various building blocks to create a scalable and efficient architecture.

Now that we have a good idea about the different high-level components our system will contain, the next thing to do is to dive deeper into the low-level design of each microservice, exploring their specific functionalities, APIs, and data flow.

Designing Tweet Service

Tweet Service is responsible for handling the creation, retrieval, and deletion of tweets. It plays a crucial role in the Twitter-like service by managing the core functionality related to tweets. Let's explore the low-level design of Tweet Service. *Figure 11.3* shows the architecture of Tweet Service where we have the appropriate requests flow in from the load balancer, through the API gateway to Tweet Service, which interacts with the **Tweet Database** table, the object store, and the message queue to power **Timeline Service** and **Search Service** (discussed in the next sections).

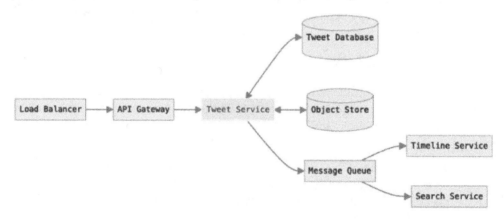

Figure 11.3: High-level architecture of Tweet Service

Each service will expose API endpoints that can be used to communicate with it. Here are the API endpoints for Tweet Service:

- POST /tweets – create a new tweet:

  - Request body: Tweet content, user ID, and optional media attachment

  - Response: Created tweet object with assigned tweet ID and timestamp

- GET /tweets/{tweetId} – retrieve a specific tweet by its ID:

  - Response: Tweet object containing content, user details, timestamp, and engagement metrics (e.g., likes, retweets)

- DELETE /tweets/{tweetId} – delete a tweet by its ID:

  - Request: User authentication token to ensure only the tweet owner can delete it

  - Response: Success or error message

- GET /users/{userId}/tweets – retrieve tweets posted by a specific user:

  - Request: User ID and optional pagination parameters

  - Response: List of tweet objects posted by the user

These APIs are sufficient for any Tweet Service client to interact with the service. Let us now understand the data model and storage aspects of this service.

Data storage

Tweet Service will utilize a combination of database and object storage for storing tweet data:

- **Database (e.g., Apache Cassandra or Amazon DynamoDB)**: Structured tweet data such as tweetId, userId, content, and timestamp can be stored in a relational database. The partition key could be based on tweetId to ensure an even distribution of data across nodes, and the clustering key could be a timestamp for the efficient retrieval of tweets in chronological order.

- **Object storage (e.g., Amazon S3)**: Tweets are sometimes posted with media attachments that can be stored as separate objects in object storage. Each such file will be assigned a unique identifier and tweets in the database will contain a reference to the media file's unique identifier.

Given that we have discussed the APIs and the data storage aspects of the service, let us discuss the tweet creation and retrieval flows.

Tweet creation flow

Figure 11.4 shows the flow of data/API calls invoked in the tweet creation flow.

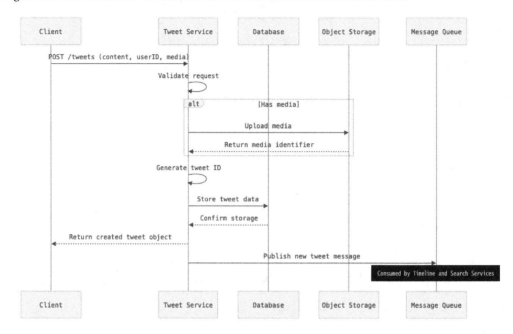

Figure 11.4: Tweet creation flow

When a user creates a new tweet through the client application, the following occurs:

1. The **client** sends a POST request to the /tweets endpoint with the tweet content, user ID, and optional media attachment.

2. **Tweet Service** receives the request and performs the necessary validations (e.g., tweet length and user authentication).

3. If the tweet contains a media attachment, **Tweet Service** uploads the media file to the object storage and obtains the unique identifier.

4. **Tweet Service** generates a unique tweet ID and stores the tweet data (content, user ID, timestamp, and media reference) in the database.

5. The created tweet object is returned to the client as the response.

6. **Tweet Service** publishes a message to the message queue (e.g., Apache Kafka) containing the newly created tweet ID. This message will be consumed by other services, such as Timeline Service and Search Service, for further processing.

Tweet retrieval flow

Figure 11.5 shows the data flow/API calls invoked in the tweet retrieval flow.

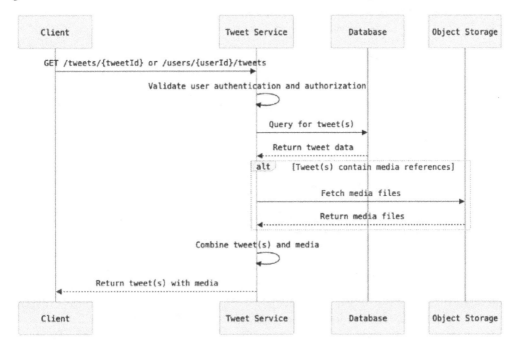

Figure 11.5: Tweet retrieval flow

When a user requests to view a specific tweet or a user's tweet timeline, the following steps are carried out:

1. The client sends a GET request to the appropriate endpoint (/tweets/{tweetId} or /users/{userId}/tweets) with the necessary parameters.

2. **Tweet Service** receives the request and validates the user authentication and authorization.

3. **Tweet Service** queries the database to retrieve the requested tweet(s) based on the provided tweet ID or user ID.

4. If the tweet(s) contain media references, **Tweet Service** fetches the corresponding media files from the object storage.

5. The retrieved tweet(s) and media files are combined and returned to the client as a response.

For any system designed to be used by a large number of users and services, it is important to build a layer of caching to improve performance, which we will cover next.

Caching

To improve the performance of tweet retrieval, Tweet Service can utilize a distributed caching layer (e.g., Redis). Frequently accessed tweets, such as popular or trending tweets, can be cached in the cache layer. When a tweet is requested, Tweet Service first checks the cache. If the tweet is found in the cache, it is served directly from there, reducing the load on the database. If the tweet is not found in the cache, Tweet Service retrieves it from the database, stores it in the cache for future requests, and returns it to the client.

The following are some of the caching strategies for new and popular tweets:

- Time-based sliding window cache:

  - Maintain a cache of tweets posted within the last N hours (e.g., 24 hours)

  - As new tweets are posted, add them to the cache and remove tweets older than the time window

  - This ensures that the newest tweets are always available in the cache

- Popularity-based caching:

  - Implement a scoring system based on engagement metrics (likes, retweets, and replies)

  - Cache tweets with scores above a certain threshold

  - Periodically recalculate scores and update the cache accordingly

- Hybrid approach:

  - Combine time-based and popularity-based strategies

  - Cache all tweets from the last few hours (e.g., two hours) regardless of popularity

  - For older tweets, only cache those that meet a popularity threshold

- Predictive caching:

  - Use machine learning models to predict which tweets are likely to become popular

  - Proactively cache tweets that the model predicts will have high engagement

- User-based caching:

  - Cache recent tweets from users with high follower counts or verified status

  - This strategy assumes that tweets from popular users are more likely to be requested

The following are the eviction strategies:

- **Least Recently Used (LRU)**:

  - Evict the least recently accessed tweets when the cache reaches capacity

  - This strategy works well for maintaining a cache of currently popular content

- **Time to Live (TTL)**:

  - Assign an expiration time to each cached tweet

  - Evict tweets that have exceeded their TTL

  - Use shorter TTLs for regular tweets and longer TTLs for highly popular tweets

- **Least Frequently Used (LFU)**:

  - Track the access frequency of cached tweets

  - Evict the least frequently accessed tweets when the cache is full

  - This can be combined with a decay factor to favor more recent popularity

- **Size-based eviction**:

  - Implement a maximum cache size (e.g., 10 GB)

  - When the cache reaches this limit, evict tweets based on a combination of size and another factor (such as LRU or LFU)

- **Priority-based eviction**:

  - Assign priorities to tweets based on factors such as user popularity, tweet engagement, and recency

  - Evict lower-priority tweets first when the cache is full

The following are the implementation considerations:

- Use a multi-tiered caching strategy, such as hot cache for extremely popular tweets, warm cache for moderately popular tweets, and cold storage for less accessed tweets

- Implement cache warming techniques to preload the cache with likely-to-be-accessed tweets after a system restart

- Use cache versioning or generation numbers to handle cache invalidation when tweets are modified or deleted

- Consider using separate caches for different types of data (e.g., tweet content, user profiles, and timelines) to optimize performance and eviction strategies for each

- Implement monitoring and analytics to continuously evaluate and refine caching strategies based on actual usage patterns

By implementing these caching and eviction strategies, the Twitter-like service can efficiently manage its cache to keep the most relevant and frequently accessed tweets readily available, significantly improving response times and reducing the load on backend databases.

By following this low-level design, Tweet Service can efficiently handle the creation, retrieval, and deletion of tweets while ensuring scalability, performance, and data integrity. The service integrates with other components, such as the object storage and message queue, to provide a seamless tweet management experience.

In the subsequent sections, we will explore the low-level design of other critical services, such as User Service, Timeline Service, and Search Service, which work in conjunction with Tweet Service to power the Twitter-like platform.

Designing User Service

User Service is responsible for managing user-related functionalities in the Twitter-like service. It handles user registration, authentication, profile management, and follower-followee relationships. Let's dive into the low-level design of User Service. *Figure 11.6* shows the User Service high-level architecture, where we have the client request flowing through the load balancer, API gateway, and User Service and interacting with the user and follow tables.

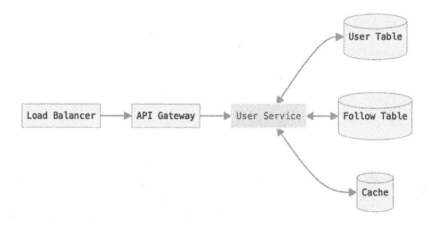

Figure 11.6: High-level architecture of User Service

Each service is defined by the set of API endpoints it exposes to its clients and how it stores the data. We will cover both these aspects next.

User Service will expose the following API endpoints:

- `POST /users` – create a new user account:
    - `Request body`: User information, such as username, email, and password
    - `Response`: Created user object with assigned user ID
- `GET /users/{userId}` – retrieve user profile information:
    - `Response`: User object containing profile details, such as username, bio, profile picture URL, and follower/followee counts
- `PUT /users/{userId}` – update user profile information:
    - `Request body`: Updated user information, such as bio, profile picture URL, or location
    - `Response`: Updated user object
- POST /users/{userId}/follow – follow another user:
    - `Request`: User authentication token and target user ID
    - `Response`: Success or error message
- DELETE /users/{userId}/follow – unfollow another user:
    - `Request`: User authentication token and target user ID
    - `Response`: Success or error message
- `GET /users/{userId}/followers` – retrieve a user's followers:
    - `Response`: List of user objects representing the followers
- `GET /users/{userId}/following` – retrieve the users a user is following:
    - `Response`: List of user objects representing the followees

Next, we will cover how and where the data for User Service will be stored.

Data storage

User Service will store user-related data in a database (e.g., PostgreSQL or MySQL):

- **User table**: This table is used to store user information, such as user ID, username, email, password hash, bio, profile picture URL, location, and registration timestamp, and will use the user ID as the primary key for efficient lookup

- **Follow table**: This table will store the follower-followee relationships. It will contain the follower user ID, followee user ID, and timestamp columns. This table will use a composite primary key of the follower user ID and followee user ID to ensure uniqueness and efficient querying.

Now that we have covered the API endpoints and the data storage, let us look at the flow for user registration, authentication, follows, and retrieving followers.

User registration flow

Figure 11.7's sequence diagram shows the process of registering a new user, including validation, user ID generation, password hashing, and data storage.

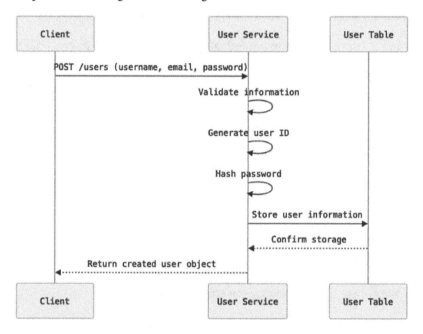

Figure 11.7: User registration flow

When a new user registers for the Twitter-like service:

1. The **client** sends a POST request to the /users endpoint with the user's information, such as username, email, and password.

2. **User Service** receives the request and performs the necessary validations (e.g., has a unique username and email been provided?).

3. If the validations pass, **User Service** generates a unique user ID and stores the user information in the user table, hashing the password for security.

4. The created user object is returned to the client as the response.

User authentication flow

Figure 11.8's diagram illustrates the process of user authentication, including credential verification and token generation.

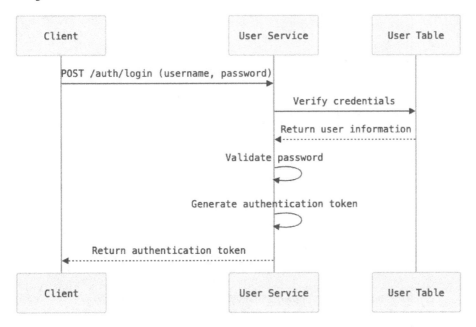

Figure 11.8: Authentication flow

When a user logs in to the Twitter-like service:

1. The **client** sends a POST request to the authentication endpoint (e.g., /auth/login) with the user's credentials (e.g., username and password).

2. **User Service** receives the request and verifies the provided credentials against the stored user information in the user table.

3. If the credentials are valid, **User Service** generates an authentication token (e.g., **JSON Web Token**, or **JWT**) containing the user ID and other relevant information.

4. The authentication token is returned to the client, which includes it in subsequent requests to authenticate and authorize the user.

Follow/unfollow flow

Figure 11.9's sequence diagram shows the process of following or unfollowing a user, including token verification and database updates.

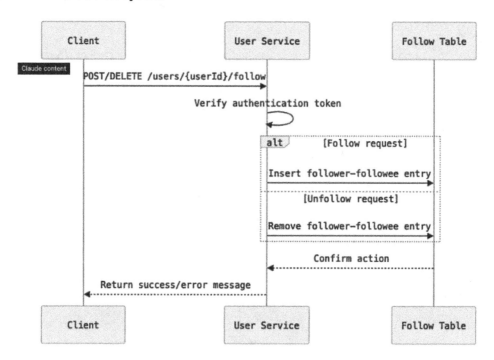

Figure 11.9: Follow/unfollow flow sequence diagram

When a user follows or unfollows another user:

1. The **client** sends a POST or DELETE request to the /users/{userId}/follow endpoint with the user authentication token and target user ID.

2. **User Service** verifies the authentication token to ensure the request is made by a valid user.

3. For a follow request, **User Service** inserts a new entry into the follow table with the follower user ID and followee user ID.

4. For an unfollow request, **User Service** removes the corresponding entry from the follow table.

5. **User Service** returns a success or error message to the client.

Retrieving followers/followees

Figure 11.10 illustrates the process of retrieving a user's followers or followees, including database queries and profile fetching.

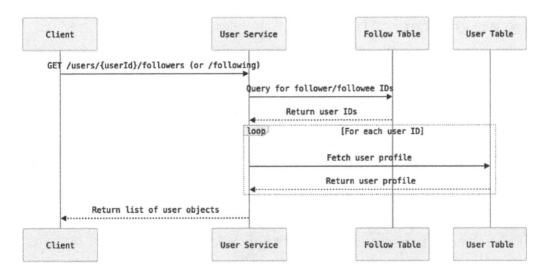

Figure 11.10: Retrieving followers/followees sequence diagram

When a user requests their followers or the users they are following, the following occurs:

1. The **client** sends a GET request to the /users/{userId}/followers or /users/ {userId}/following endpoint.

2. **User Service** queries the follow table to retrieve the user IDs of the followers or followees.

3. **User Service** then fetches the user profile information for each follower or followee from the user table.

4. The list of user objects representing the followers or followees is returned to the client.

By following this low-level design, User Service can efficiently handle user registration, authentication, profile management, and follower-followee relationships. It integrates with the database and caching layer to provide a seamless user experience and support the social networking aspects of the Twitter-like service.

In the next section, we will explore the low-level design of Timeline Service, which is responsible for generating and serving user timelines based on the tweets from the users they follow.

Low-level design – Timeline Service

Timeline Service is responsible for generating and serving user timelines in the Twitter-like service. It aggregates tweets from the users a person follows and presents them in chronological order. Let's explore the low-level design of Timeline Service. We have already seen in *Figure 11.2* how Tweet Service posts the incoming tweet to Timeline Service via a message queue (Kafka).

Timeline Service will expose the following API endpoints:

- `GET /timeline/{userId}` – retrieve the user's home timeline:
- Request: User authentication token
- Response: List of tweet objects representing the user's timeline
- `GET /timeline/{userId}/mentions` – retrieve the user's mentions timeline:
  - Request: User authentication token
  - Response: List of tweet objects mentioning the user

Given that we have the API endpoints listed, we will now check the data flow.

Data flow

Timeline Service relies on Tweet Service and User Service to generate user timelines. *Figure 11.11* shows the data flow for creating a new tweet.

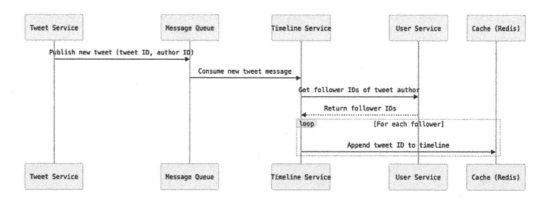

Figure 11.11: The data flow for creating a new tweet and updating followers' timelines

Here is the flow:

1. When a new tweet is created, **Tweet Service** publishes a message to the message queue (e.g., Apache Kafka) containing the tweet ID and the user ID of the tweet author.

2. **Timeline Service** consumes the message from the message queue.

3. **Timeline Service** retrieves the follower IDs of the tweet author from User Service.

4. For each follower, **Timeline Service** appends the tweet ID to their timeline data structure (e.g., a list or a sorted set) stored in the cache (e.g., Redis).

5. The timeline data structure maintains a limited number of recent tweet IDs for each user, typically based on a time window or a maximum count.

Timeline retrieval flow

Figure 11.12 shows the timeline retrieval flow.

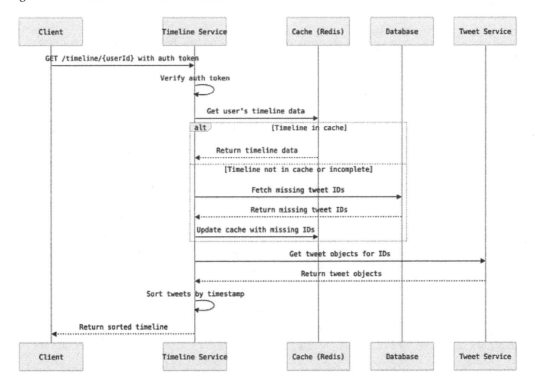

Figure 11.12: The timeline retrieval flow when a user requests their home timeline

When a user requests their home timeline, the following occurs:

- The **client** sends a GET request to the `/timeline/{userId}` endpoint with the user authentication token.

- **Timeline Service** verifies the authentication token to ensure the request is made by a valid user.

- **Timeline Service** retrieves the user's timeline data structure from the cache.

- If the timeline data structure is not found in the cache or is incomplete, **Timeline Service** fetches the missing tweet IDs from the database and updates the cache accordingly.

- **Timeline Service** retrieves the actual tweet objects corresponding to the tweet IDs from **Tweet Service**.

- The retrieved tweets are sorted based on their timestamps to ensure chronological order.

- The sorted list of tweet objects is returned to the client as the user's home timeline.

Mentions timeline

The mentions timeline is generated similarly to the home timeline but with a different data flow. *Figure 11.13* shows the mentions timeline process for handling and retrieving mentions.

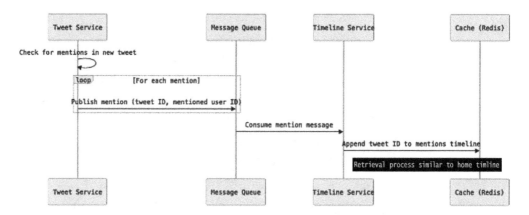

Figure 11.13: The mentions timeline process for handling and retrieving mentions

Here is the flow:

1. When a new tweet is created, **Tweet Service** checks whether the tweet contains any mentions of other users.

2. If mentions are found, **Tweet Service** publishes a separate message to the message queue for each mentioned user, containing the tweet ID and the mentioned user ID.

3. **Timeline Service** consumes these messages and appends the tweet ID to the mentions timeline data structure of each mentioned user in the cache.

4. The mentions timeline retrieval process is similar to the home timeline retrieval, but uses the mentions timeline data structure instead.

Push-based updates

To provide real-time updates to user timelines, Timeline Service can utilize push-based mechanisms. When a new tweet is created, Timeline Service can send a real-time notification to the relevant users' clients using WebSocket connections. The clients can update their timelines instantly upon receiving the notification, providing a real-time user experience.

By following this low-level design, Timeline Service can efficiently generate and serve user timelines by aggregating tweets from followed users. It leverages caching and push-based updates to provide a real-time and responsive user experience. The service integrates with Tweet Service, User Service, and the message queue to ensure data consistency and scalability.

In the next section, we will explore the low-level design of Search Service, which enables users to search for tweets and user profiles based on keywords and other criteria.

Designing Search Service

Search Service is responsible for enabling users to search for tweets and user profiles based on keywords, hashtags, and other criteria in the Twitter-like service. It provides a powerful and efficient search functionality to help users discover relevant content. Let's dive into the low-level design of Search Service. We have already seen in *Figure 11.3* how **Tweet Service** posts the incoming tweets to the message queue, which forwards it to **Search Service** for index searching.

Figure 11.14 illustrates how Search Service interacts with the API gateway, message queue, and Elasticsearch. It also shows how **Tweet Service** and **User Service** feed data into the message queue for indexing.

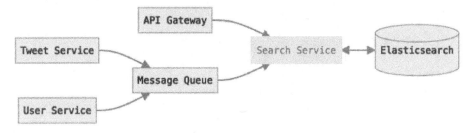

Figure 11.14: Search Service low-level design

Let us look at the API endpoints exposed by Search Service and the data flow.

The Search Service will expose the following API endpoints:

- `GET/search/tweets?q={query}&limit={limit}&offset={offset}` – search for tweets based on a given query:

- `Request`: Search query; optional limit and offset parameters for pagination

- `Response`: List of tweet objects matching the search query

- `GET/search/users?q={query}&limit={limit}&offset={offset}` – search for user profiles based on a given query:

- `Request`: Search query; optional limit and offset parameters for pagination

- `Response`: List of user objects matching the search query

We have covered the API endpoints. Let us now look at the data flow and indexing next.

Data flow and indexing

To enable efficient searching, Search Service relies on a search engine such as Elasticsearch to index and store the tweet and user data. *Figure 11.15*'s sequence diagram shows the process of indexing new tweets and user profiles in Elasticsearch, including data extraction and processing by **Search Service**.

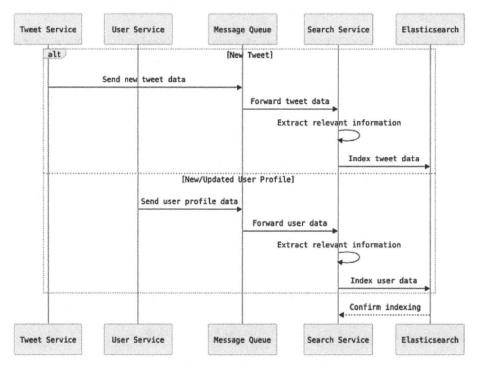

Figure 11.15: Data flow and indexing sequence diagram

Here is the flow:

1. When a new tweet is created, Tweet Service sends the tweet data to Search Service.

2. Search Service extracts relevant information from the tweet, such as the text content, hashtags, mentions, and user details.

3. The extracted data is then indexed in Elasticsearch, creating an inverted index that maps terms to tweet IDs.

4. Similarly, when a new user is created or their profile is updated, User Service sends the user data to Search Service for indexing.

5. The user data, including the username, bio, location, and other relevant information, is indexed in Elasticsearch.

We have covered the indexing of data into our Search Service. Let us now learn how to process search queries.

Search query processing

Figure 11.16 illustrates the process of handling a search query, from the initial client request through query parsing, Elasticsearch searching, and result processing.

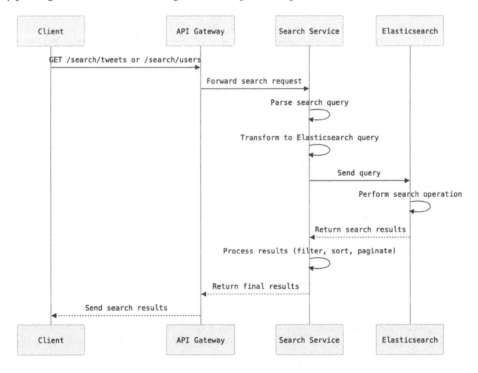

Figure 11.16: Search query processing sequence diagram

When a user performs a search query, the following occurs:

1. The client sends a GET request to the `/search/tweets` or `/search/users` endpoint with the search query and optional pagination parameters.

2. Search Service receives the request and parses the search query.

3. The parsed query is transformed into an Elasticsearch query using appropriate query builders and filters.

4. Search Service sends the query to Elasticsearch, which performs the search operation on the indexed data.

5. Elasticsearch retrieves the matching tweet or user documents based on the query criteria.

6. Search Service processes the search results, applying any additional filtering, sorting, or pagination as needed.

7. The final list of tweets or user objects is returned to the client as the search results.

Relevance scoring and ranking

To provide the most relevant search results, Search Service utilizes Elasticsearch's relevance scoring and ranking capabilities. **Elasticsearch** uses a combination of factors, such as **term frequency, inverse document frequency (TF-IDF)** and field-level boosting, to calculate the relevance score of each document. The relevance score determines the order in which the search results are presented to the user. Search Service can customize the relevance scoring by defining custom scoring functions or boosting certain fields based on specific requirements.

By following this low-level design, Search Service can provide a powerful and efficient search functionality for the Twitter-like service. It leverages Elasticsearch for indexing and searching, applies relevance scoring and ranking techniques, and utilizes caching to optimize performance. The service integrates with Tweet Service and User Service to ensure data consistency and real-time updates.

In the next section, we will discuss additional considerations and best practices for designing and implementing the Twitter-like service.

Additional considerations

When designing and implementing the Twitter-like service, there are several additional considerations and best practices to keep in mind. These considerations ensure the system is scalable, maintainable, and aligned with business requirements. Let's explore some of these key aspects:

- **Handling trending topics and hashtags**: We can implement a mechanism to track and identify trending topics and hashtags based on their popularity and frequency of use. To do so, we can utilize real-time stream processing frameworks such as Apache Storm or Apache Flink to analyze the incoming tweet data and update the trending topics in real-time. For performance, we can store the trending topics and hashtags in a cache or database for quick retrieval and display to users. Finally, we can provide API endpoints to retrieve the current trending topics and allow users to explore tweets related to those topics.

- **Implementing rate limiting and throttling**: One of the core requirements for any large-scale system with millions of users is to implement rate limiting and throttling mechanisms to protect the system from abuse and ensure fair usage of resources. We can set appropriate rate limits for different API endpoints based on the expected usage patterns and system capacity and use techniques such as token buckets or leaky buckets to enforce rate limits and throttle requests that exceed the defined thresholds.

By considering these additional aspects and best practices, the Twitter-like service can be designed and implemented to be scalable, maintainable, and extensible. It ensures a robust and user-friendly platform that can handle the demands of a growing user base and evolving business requirements.

In the final section, we will summarize the key takeaways from this chapter and discuss future considerations and potential enhancements for the Twitter-like service.

Summary

In this chapter, we have explored the system design of a Twitter-like service, covering functional and non-functional requirements, data modeling, scalability considerations, and architectural components. We examined both high-level and low-level designs, focusing on key services such as Tweet Service, User Service, Timeline Service, and Search Service. The system architecture leverages horizontal scaling, data partitioning, and distributed processing to handle a large number of users and interactions efficiently.

Throughout our discussion, we emphasized scalability, reliability, and performance, incorporating techniques such as caching and asynchronous processing to improve response times and handle high throughput. As we conclude, it's important to note that designing such a service is an ongoing process, requiring flexibility and adaptability to evolving user needs and technological advancements. By following the principles and best practices outlined here, developers and system architects can create a robust and scalable platform that meets user needs and stands the test of time. In the next chapter, we will see how we can design a service like Instagram for millions of users. Stay tuned.

12

Designing a Service Like Instagram

In the era of social media, photo-sharing platforms have taken the world by storm, and Instagram stands out as one of the most popular and influential services. With over a billion active users, Instagram has revolutionized the way people capture, share, and engage with visual content.

Designing a service similar to Instagram presents a unique set of challenges and opportunities. It requires a robust and scalable architecture that can handle the immense volume of user-generated content while providing a seamless and engaging user experience. In this chapter, we will explore the system design of an Instagram-like service, delving into the key components, design decisions, and best practices involved in building a scalable and efficient photo-sharing platform. By the end of this chapter, you will have a comprehensive understanding of the system design principles and techniques involved in building a service similar to Instagram.

In this chapter, we will cover the following topics:

- Functional requirements
- Non-functional requirements
- Designing the data model
- Scale calculations
- High-level design
- Low-level design
- Additional considerations

Let us start with the functional requirements of a service similar to Instagram.

Functional requirements

Before diving into the system design, it is crucial to define the functional requirements that specify what the Instagram-like service should be capable of doing. These requirements lay the foundation for the entire design process and ensure that the system meets the needs of its users. Let's explore the key functional requirements:

- **User registration and authentication**:

  - Users should be able to create new accounts by providing the necessary information such as a username, email, and password

  - The system should securely store user credentials and authenticate users upon login

  - User sessions should be managed efficiently to allow seamless access to the service

- **Photo upload and sharing**:

  - Users should be able to upload photos from their devices or capture photos directly within the app

  - The system should support various photo formats (e.g., JPEG and PNG) and perform necessary processing and compression

  - Users should have the ability to apply filters, add captions, and tag other users in their photos

  - The uploaded photos should be associated with the user's profile and stored in a scalable and reliable storage system

- **News feed**:

  - The news feed is a crucial feature that displays a personalized stream of photos from the users a person follows

  - The system should generate and serve the news feed in real-time, considering factors such as photo timestamps, user preferences, and engagement metrics

  - The news feed should support infinite scrolling, allowing users to load more photos as they scroll down

- **User interactions**:

  - Users should be able to like and comment on photos shared by other users

  - The system should store and display the count of likes and comments for each photo

  - Users should have the ability to mention other users in comments using the @ symbol followed by the username

- **Direct messaging**:

    - The service should support direct messaging functionality, allowing users to privately send photos and messages to other users or groups

    - Users should be able to initiate conversations, view message history, and receive real-time notifications for new messages

- **Search and discovery**:

    - Users should be able to search for other users, photos, and hashtags within the platform

    - The search functionality should provide relevant and accurate results based on keywords, usernames, and hashtags

    - The system should also support discovery features, such as exploring popular photos, trending hashtags, and personalized recommendations

- **Notifications**:

    - The service should send real-time notifications to users for various events, such as new followers, likes, comments, and direct messages

    - Notifications should be delivered through push notifications on mobile devices and in-app notifications

    - Users should have control over their notification preferences, allowing them to customize the types of notifications they receive

These functional requirements provide a comprehensive overview of the core features that an Instagram-like service should offer. By fulfilling these requirements, the system will enable users to seamlessly share photos, connect with others, and engage with visual content.

In the next section, we will explore the non-functional requirements that ensure the service remains scalable, reliable, and performant while meeting the functional requirements discussed so far.

Non-functional requirements

While functional requirements define what the system should do, non-functional requirements specify how the system should perform and behave. These requirements are critical to ensuring that the Instagram-like service remains scalable, available, and reliable under various conditions. Let's discuss the key non-functional requirements:

- **Scalability**:

    - The system should be designed to handle a large number of users and photos, accommodating growth and peak traffic loads

- Horizontal scalability should be achieved by adding more servers and distributing the load across them

- The architecture should allow for easy scaling of individual components, such as the Photo Upload Service or News Feed Service , independently

- **Performance**:

    - The service should provide a fast and responsive user experience, with minimal latency for key operations such as photo uploads, news feed loading, and interactions

    - The system should optimize resource utilization and employ caching mechanisms to reduce the load on backend services and improve performance

    - Asynchronous processing should be utilized for resource-intensive tasks, such as photo compression and thumbnail generation, to ensure quick response times

- **Availability**:

    - The service should be highly available, minimizing downtime and ensuring that users can access their photos and engage with the platform at all times

    - The system should be designed with redundancy and fault tolerance in mind, employing techniques including data replication, load balancing, and automatic failover

    - Regular backups and disaster recovery mechanisms should be in place to protect against data loss and ensure quick recovery in the case of failure

- **Reliability**:

    - The system should be reliable, ensuring data integrity and consistency across all components

    - Mechanisms should be implemented to handle and recover from errors gracefully, preventing data corruption or inconsistencies

    - Transactions and data consistency models should be chosen carefully to maintain the accuracy and reliability of user data, likes, comments, and other interactions

- **Usability**:

    - The system should provide features such as search, filters, and recommendations to enhance user engagement and content discovery

By addressing these non-functional requirements, the Instagram-like service can ensure a reliable, scalable, and performant user experience. It is essential to consider these requirements throughout the design process and make architectural decisions that align with these goals.

In the next section, we will explore the data model that forms the foundation of the Instagram-like service, defining the entities and relationships necessary to support the functional requirements.

Designing the data model

The data model is a crucial component of the Instagram-like service, as it defines the structure and relationships of the data entities involved. A well-designed data model ensures efficient storage, retrieval, and manipulation of data while supporting the functional requirements of the system. Let's dive into the key entities and their relationships:

- The User entity represents the users of the Instagram-like service, storing their profile information such as username, email, password hash, profile picture, bio, and website. Each user is uniquely identified by their user_id. The Photo entity represents the photos uploaded by users. It contains the photo details such as the user_id of the uploader, caption, image URL, creation timestamp, location, and tags. The photo_id serves as the primary key.

- The Comment entity represents the comments posted on photos. It contains the comment text, user_id of the commenter, photo_id of the associated photo, and creation timestamp. The comment_id is the primary key. The Like entity represents the likes on photos. It contains the user_id of the user who liked the photo and the photo_id of the liked photo. The created_at timestamp records when the like occurred.

- The Follow entity represents the follow relationships between users. It contains the follower_id and followee_id, indicating which user is following whom. The created_at timestamp captures when the follow relationship was established. The Hashtag entity represents the hashtags used in photos. It contains the hashtag_id and the name of the hashtag. The PhotoHashtag entity represents the many-to-many relationship between photos and hashtags. It contains the photo_id and hashtag_id, indicating which photos are associated with which hashtags. The DirectMessage entity represents private conversations between users. It contains the message_id, sender_id, recipient_id, content, and created_at timestamp.

Figure 12.1 illustrates the key entities (User, Photo, Comment, Like, Follow, Hashtag, PhotoHashtag, and DirectMessage) in the Instagram-like service, their attributes, and the relationships between them, providing a comprehensive visual representation of the data structure that supports the service's core functionalities.

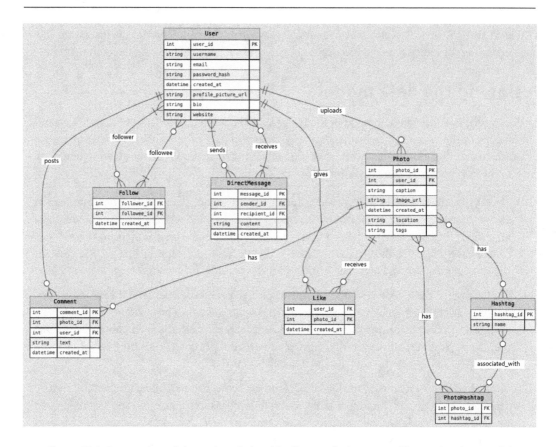

Figure 12.1: An overview of the entity relationship diagram for Instagram-like service data model

This diagram effectively captures the complex relationships between the various entities in the Instagram-like service, including one-to-many and many-to-many relationships. It visually represents how users interact with photos, comments, likes, follows, hashtags, and direct messages, which is crucial for understanding the data structure of the system.

By designing the data model with these entities and relationships, the Instagram-like service can efficiently store and retrieve data related to users, photos, comments, likes, follows, hashtags, and direct messages. The data model supports the functional requirements and enables the system to handle the complex interactions between users and their shared content.

After we have evaluated the data model design, we will perform scale calculations to estimate the storage, bandwidth, and processing requirements of the Instagram-like service based on the anticipated user base and usage patterns.

Scale calculations

To design a scalable Instagram-like service, it is essential to estimate the storage, bandwidth, and processing requirements based on the expected user base and usage patterns. These calculations help in making informed decisions about the infrastructure and resources needed to support the service. Let's perform some scale calculations:

- **Assumptions**:

  - Total number of users: 100 million

  - Daily active users: 10 million

  - Average number of photos uploaded per user per day: 2

  - Average photo size: 5 MB

  - Retention period for photos: 5 years

  - Average number of followers per user: 500

  - Average number of likes per photo: 100

  - Average number of comments per photo: 10

- **Storage requirements**:

  - Daily photo storage: `10 million users × 2 photos/user/day × 5 MB/photo = 100 TB/day`

  - Yearly photo storage: `100 TB/day × 365 days = 36.5 PB/year`

  - Total photo storage for 5 years: `36.5 PB/year × 5 years = 182.5 PB`

  - User data storage: `Assuming 1 MB of storage per user for profile information`

  - Total user data storage: `100 million users × 1 MB/user = 100 GB`

  - Metadata storage (likes, comments, hashtags): `Assuming 1 KB of metadata per photo`

  - Daily metadata storage: `10 million users × 2 photos/user/day × 1 KB/photo = 20 GB/day`

  - Yearly metadata storage: `20 GB/day × 365 days = 7.3 TB/year`

  - Total metadata storage for 5 years: `7.3 TB/year × 5 years = 36.5 TB`

  - Total storage: `182.5 PB (photos) + 100 GB (user data) + 36.5 TB (metadata) ≈ 182.5 PB`

- **Bandwidth requirements**:

  - Daily bandwidth for photo uploads: `10 million users × 2 photos/user/day × 5 MB/photo = 100 TB/day`

  - Daily bandwidth for photo delivery: `10 million users × 500 followers/user × 2 photos/user/day × 5 MB/photo = 50 PB/day`

  - Total daily bandwidth: `100 TB (uploads) + 50 PB (delivery) ≈ 50 PB/day`

- **Processing requirements**:

  - Peak photo uploads per second: `10 million users × 2 photos/user/day ÷ 86400 seconds/day ≈ 230 photos/second`

  - Likes per second: `230 photos/second × 100 likes/photo = 23,000 likes/second`

  - Comments per second: `230 photos/second × 10 comments/photo = 2,300 comments/second`

These calculations provide a rough estimate of the storage, bandwidth, and processing requirements for the Instagram-like service. It's important to note that these numbers can vary based on the actual usage patterns and growth of the user base. The infrastructure should be designed to handle peak loads and should be easily scalable to accommodate future growth.

In the next section, we will propose a high-level design architecture that takes into account these scale requirements and leverages various building blocks to create a scalable and efficient system.

High-level design

Now that we have a clear understanding of the functional and non-functional requirements, as well as the scale calculations, let's dive into the high-level design of the Instagram-like service. The goal is to create an architecture that is scalable, reliable, and efficient in handling the massive volume of photos, users, and interactions. *Figure 12.2* shows the high-level design of the Instagram-like system. This diagram illustrates the comprehensive architecture of an Instagram-like service, showcasing the flow of data and interactions between various components including client applications, load balancer, API gateway, microservices, databases, caching systems, object storage, CDN, and message queues, all working together to provide a scalable and efficient photo-sharing platform.

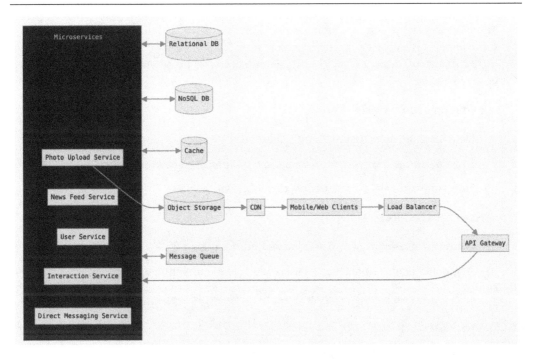

Figure 12.2: High-level system design of our Instagram-like service

Components and modules in the high-level architecture

Let's go through the following software components and modules:

- **Client-server architecture**: The Instagram-like service will follow a client-server architecture, where clients (such as mobile apps or web browsers) communicate with the server-side components through APIs. The server-side components will handle the core functionality, data storage, and processing.

- **Load balancer**: To distribute the incoming traffic evenly across multiple servers, a load balancer will be placed in front of the server-side components. The load balancer will ensure that requests are efficiently routed to the appropriate servers based on factors such as server load, request type, and geographic location.

- **API gateway**: An API gateway will act as the entry point for all client requests. It will handle request routing, authentication, rate limiting, and request/response transformation. The API gateway will expose well-defined APIs for various functionalities, such as photo uploads, news feed retrieval, user interactions, and direct messaging.

- **Microservices architecture**: To promote modularity, scalability, and maintainability, the server-side components will be designed as microservices. Each microservice will be responsible for a specific domain or functionality. The main microservices in the Instagram-like service will be the following:

  - **Photo Upload Service** :

    - Handles photo uploads, processing, and storage
    - Performs image compression, resizing, and thumbnail generation
    - Stores photos in an object storage system (e.g., Amazon S3)

  - **News Feed Service** :

    - Generates and serves personalized news feeds for users
    - Aggregates photos from followed users and applies ranking algorithms
    - Utilizes caching to improve performance and reduce latency

  - **User Service**:

    - Manages user profiles, authentication, and authorization
    - Handles user-related operations such as sign-up, login, and profile updates
    - Stores user data in a database (e.g., MySQL or PostgreSQL)

  - **Interaction service**:

    - Handles user interactions such as likes, comments, and hashtags
    - Stores interaction data in a database and updates relevant counts
    - Sends notifications to users for new interactions

  - **Direct Messaging service**:

    - Enables private communication between users
    - Handles message delivery, storage, and real-time updates
    - Utilizes a message queue (e.g., Apache Kafka) for asynchronous processing

- **Database and caching**: The Instagram-like service will employ a combination of databases and caching to store and retrieve data efficiently:

  - Relational database (e.g., MySQL or PostgreSQL):

    - Stores structured data such as user profiles, photo metadata, comments, likes, and hashtags

- Provides ACID properties and supports complex queries

- NoSQL database (e.g., Apache Cassandra or MongoDB):

  - Handles high-volume write operations for user interactions and real-time data

  - Offers scalability and eventual consistency

- Caching (e.g., Redis or Memcached):

  - Caches frequently accessed data such as news feed photos, user profiles, and trending hashtags

  - Reduces the load on databases and improves response times

- **Object storage**: Photos and videos will be stored in an object storage system (e.g., Amazon S3) that provides scalable and durable storage. Object storage allows for efficient retrieval and delivery of media files to users.

- **Content Delivery Network (CDN)**: To optimize the delivery of photos and videos to users across different geographic locations, a CDN (e.g., Amazon CloudFront) will be utilized. The CDN caches and serves content from edge locations closer to the users, reducing latency and improving the overall user experience.

- **Asynchronous processing**: Resource-intensive tasks, such as photo processing and notifications, will be handled asynchronously using message queues (e.g., Apache Kafka) and background workers. This ensures that the main application remains responsive and can handle a high volume of requests.

- **Security and privacy**: Security and privacy measures will be implemented throughout the system. User authentication and authorization will be handled using secure protocols such as OAuth. Sensitive data, such as passwords, will be hashed and stored securely. Data encryption will be applied to protect user information both in transit and at rest. Rate limiting and throttling mechanisms will be put in place to prevent abuse and ensure fair usage of the service.

This high-level design provides an overview of the key components and their interactions in the Instagram-like service. It takes into account the scale requirements and employs various building blocks to create a scalable and efficient architecture.

In the next sections, we will explore the low-level design of key components, including the Photo Upload Service , News Feed Service , and User Service, diving deeper into their specific functionalities and design considerations.

Low-level design

In this section, we will cover the low-level design of a subset of services in our system, namely the Photo Upload Service , News Feed Service, and the User Service. Let us first discuss the Photo Upload Service .

Designing the Photo Upload Service

The **Photo Upload Service** is a critical component of the Instagram-like service, responsible for handling the upload, processing, and storage of photos. Let's explore the low-level design of the Photo Upload Service in detail. *Figure 12.3* illustrates the architecture of the Photo Upload Service showing its interactions with the database, object storage, message queue, background workers, and CDN for efficient photo handling and delivery.

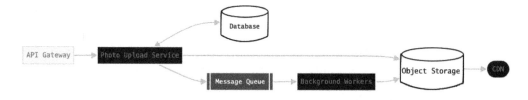

Figure 12.3: Low-level design of the Photo Upload Service

Let us now look into the API endpoint exposed by the service and the photo upload and retrieval flow.

The **Photo Upload Service** will expose the following API endpoints:

- `POST /photos`: Upload a new photo
- `Request body`: Photo file, user ID, caption, location, and other metadata
- `Response`: Uploaded photo details, including photo ID and URL

Let us look at the photo upload flow and photo retrieval flow in detail.

Photo upload flow

The following sequence diagram depicts the flow of uploading a photo, from the initial client request through metadata storage, photo upload, and asynchronous processing.

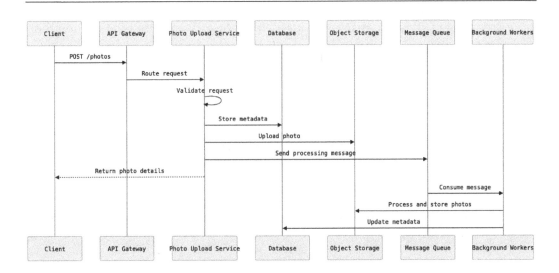

Figure 12.4: Flowchart for the photo upload flow

When a user uploads a photo through the client application, the following happens:

1. The **Client** sends a POST request to the /photos endpoint with the photo file and associated metadata.

2. The **API gateway** receives the request and routes it to the Photo Upload Service.

3. The **Photo Upload Service** performs the following steps:

    I. Validates the request, ensuring the presence of required fields and the authenticity of the user.

    II. Generates a unique photo ID and stores the metadata in the database.

    III. Uploads the photo file to the object storage system (e.g., Amazon S3).

    IV. Initiates asynchronous processing tasks for photo compression, resizing, and thumbnail generation.

    V. Returns the uploaded photo details, including the photo ID and URL, to the client.

The **client** receives the response and updates the user interface accordingly.

To optimize the photo upload process and ensure a responsive user experience, photo processing tasks are performed asynchronously:

1. After the photo is uploaded to the object storage, the **Photo Upload Service** sends a message to a message queue (e.g., Apache Kafka) containing the photo ID and processing instructions.

2. Background workers consume the messages from the message queue and perform the following tasks:

    I.   Compress the photo to optimize storage and network usage.

    II.   Generate multiple resized versions of the photo for different device resolutions.

    III. Create thumbnail images for preview purposes.

    IV. Update the photo metadata in the database with the processed file locations.

3. The processed photo files are stored in the object storage system, and their URLs are updated in the database.

We have looked at the flow for photo upload, now let us understand what happens when a user requests to view the photo.

Photo retrieval

The following sequence diagram shows the photo retrieval process, including metadata caching and content delivery through the CDN, illustrating how the system optimizes photo access for users.

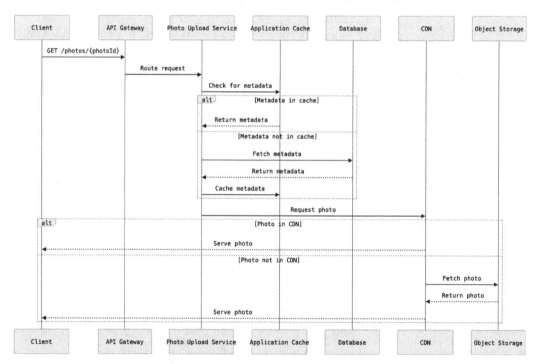

Figure 12.5: Photo retrieval flow

When a user requests to view a photo, the following occurs:

1. The **client** sends a `GET` request to the `/photos/{photoId}` endpoint.

2. The **API Gateway** receives the request and routes it to the **Photo Upload Service** .

3. The **Photo Upload Service** retrieves the photo metadata from the database, including the photo URL.

4. If the photo is not cached in the **CDN**, the **Photo Upload Service** fetches the photo file from the object storage system.

5. The photo file is returned to the **client**, along with the necessary metadata.

6. The **client** renders the photo on the user interface.

Caching and content delivery

To improve photo retrieval performance and reduce the load on the backend services, caching and content delivery mechanisms are employed:

-   **CDN**: Frequently accessed photos are cached in the CDN, which serves them from edge locations closer to the users. The CDN reduces the latency and improves the photo-loading speed for users across different geographic regions.

-   **Application-level caching**: The Photo Upload Service can utilize an in-memory caching system (e.g., Redis) to store frequently accessed photo metadata. Caching photo metadata reduces the number of database queries and improves response times.

By following this low-level design, the Photo Upload Service ensures efficient and reliable handling of photo uploads, processing, and retrieval. It leverages asynchronous processing, caching, and content delivery techniques to provide a seamless and performant user experience.

In the next section, we will explore the low-level design of the News Feed Service, which is responsible for generating and serving personalized photo feeds to users.

News Feed Service

The **News Feed Service** is responsible for generating and serving personalized photo feeds to users based on their follow relationships and engagement activities. Let's dive into the low-level design of the News Feed Service . *Figure 12.6* illustrates the architecture of the News Feed Service, showing its interactions with the database, cache, notification service, and WebSocket Service for efficient feed generation and real-time updates.

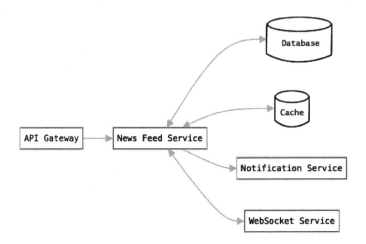

Figure 12.6: Low-level design of the News Feed Service

Let us look at the API endpoints exposed by the News Feed Service and the generation process.

The **News Feed Service** exposes the following API endpoint:

- `GET /newsfeed/{userId}`: Retrieve the personalized news feed for a user:

  - `Request parameters`: User ID, pagination token, and limit

  - `Response`: List of photo objects in the user's news feed

Let us dive into the news feed generation process in detail.

Generating news feeds

The News Feed Service generates personalized news feeds for users based on the following steps:

- **Follower-followee relationship**: The News Feed Service retrieves the list of users that the current user follows from the database. This information is stored in the `Follow` table, which contains the follower-followee relationships.

- **Photo aggregation**: For each followed user, the News Feed Service retrieves their recently uploaded photos from the database. The photo metadata, including the photo ID, user ID, timestamp, and engagement metrics (likes and comments), is fetched.

- **Ranking and ordering**: The retrieved photos are ranked and ordered based on relevance and freshness. Various ranking algorithms can be applied, such as chronological ordering, engagement-based ranking, or a combination of factors. The ranking algorithm takes into account the photo's timestamp, the user's affinity with the photo owner, and the overall engagement received by the photo.

- **Pagination**: To optimize performance and network usage, the news feed is paginated. The API endpoint accepts a pagination token and a limit to retrieve a subset of the news feed photos. The pagination token represents the last photo ID or timestamp from the previous page, allowing efficient retrieval of the next set of photos.

- **Caching**: To improve the response time and reduce the load on the backend services, the generated news feed is cached. The News Feed Service utilizes a distributed caching system (e.g., Redis) to store the news feed data. The cache is updated whenever new photos are uploaded or engagement activities occur.

- **Photo deduplication**: To prevent duplicate photos from appearing in the news feed, the News Feed Service performs deduplication. Deduplication can be achieved by maintaining a set of photo IDs that have already been included in the user's news feed. Before adding a photo to the news feed, the service checks if the photo ID exists in the deduplication set.

Real-time updates

To provide real-time updates to the news feed, the News Feed Service employs the following mechanisms:

- **WebSocket or long polling**: The client establishes a persistent connection with the server using WebSocket or long polling techniques. Whenever a new photo is uploaded by a followed user or significant engagement occurs on a photo in the user's news feed, the server pushes the update to the client in real-time. The client receives the update and dynamically updates the news feed on the user interface.

- **Notification service integration**: The News Feed Service integrates with the **Notification service** to send push notifications to users for important updates. When a followed user uploads a new photo or when a photo in the user's news feed receives significant engagement, a notification is triggered. The Notification service sends the notification to the user's device, prompting them to view the updated news feed.

Feed synchronization

To ensure a consistent news feed experience across multiple devices, the News Feed Service implements feed synchronization:

- **Timestamp-based synchronization**: Each photo in the news feed is associated with a timestamp indicating when it was added to the feed. When a user accesses their news feed from a different device, the client sends the timestamp of the last viewed photo. The News Feed Service uses this timestamp to retrieve and return the photos that have been added to the feed since the last viewed timestamp.

- **Incremental updates**: Instead of retrieving the entire news feed each time, the client can request incremental updates. The client sends the last viewed photo ID or timestamp, and the News Feed Service returns only the new or updated photos since that point. This approach reduces the amount of data transferred and improves the efficiency of feed synchronization.

By following this low-level design, the News Feed Service generates personalized and engaging photo feeds for users. It leverages ranking algorithms, caching, real-time updates, and feed synchronization techniques to provide a seamless and up-to-date news feed experience.

In the next section, we will explore the low-level design of the User Service , which manages user profiles, authentication, and social interactions within the Instagram-like service.

User Service

The **User Service** is responsible for managing user profiles, authentication, and social interactions within the Instagram-like service. It handles user registration, login, profile updates, and follow/unfollow functionality. Let's explore the low-level design of the User Service. *Figure 12.7* illustrates the architecture of the User Service, showing its interactions with the database, cache, and authentication service for managing user profiles and authentication.

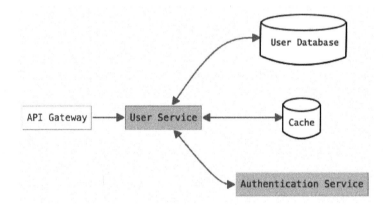

Figure 12.7: Low-level design of the User Service

Let us look at the API endpoints exposed by the User Service, next.

The User Service exposes the following API endpoints:

- POST /users: Create a new user account:

    - Request body: User information, such as username, email, password, and profile picture

    - Response: User ID and authentication token

- `POST /users/login`: Authenticate the user and generate an authentication token:

  - `Request body`: User credentials (email/username and password)

  - `Response`: Authentication token

- `GET /users/{userId}`: Retrieve user profile information:

  - `Request parameters`: User ID

  - `Response`: User profile data, including username, profile picture, bio, and follower/following counts

- `PUT /users/{userId}`: Update user profile information:

  - `Request parameters`: User ID

  - `Request body`: Updated user profile data

  - `Response`: Success status

- `POST /users/{userId}/follow`: Follow a user:

  - `Request parameters`: User ID of the user to follow

  - `Response`: Success status

- `DELETE /users/{userId}/follow`: Unfollow a user:

  - `Request parameters`: User ID of the user to unfollow

  - `Response`: Success status

Now that we have looked at the different APIs exposed by the service, let us understand the user registration and authentication flow.

User registration and authentication

The following sequence diagram depicts the process of user registration and authentication, showing the steps involved in creating a new account and logging in to an existing account.

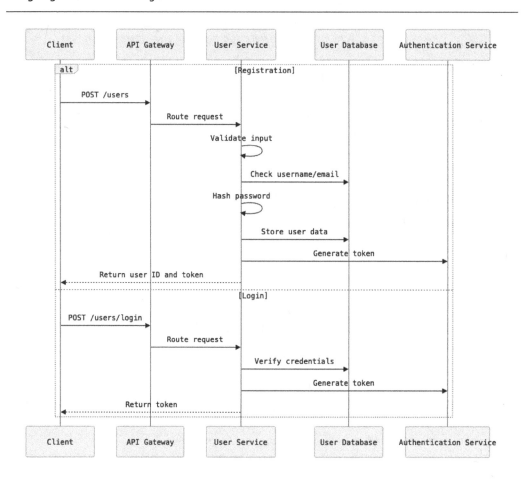

Figure 12.8: User registration and authentication flow sequence diagram

Let us understand what is happening in the preceding sequence diagram. When a user registers for the Instagram-like service or logs in, the following occurs:

1. The **client** sends a POST request to the /users or /users/login endpoints with the user's information or credentials.

2. The **User Service** receives the request and performs the following steps:

- For registration, it does the following:

  - Validates the user input data

  - Checks whether the username or email is already taken

  - Hashes the password securely

  - Generates a unique user ID

  - Stores the user information in the database

- For login, it does the following:

  - Verifies the provided credentials against the stored user data

  - If the credentials are valid, generate an authentication token (e.g., JWT)

3. The **User Service** returns the user ID and authentication token to the client.

4. The **client** stores the authentication token securely and includes it in subsequent requests to authenticate the user.

When a user updates their profile information, the client sends a PUT request to the /users/{userId} endpoint with the updated profile data. The User Service validates the request and updates the user's profile in the database and the updated profile information is returned to the client.

Let us look at the flow of events when users follow/unfollow each other in our service.

Follow/unfollow functionality

The following sequence diagram illustrates the step-by-step process and interactions between system components during the follow and unfollow operations, including request validation, database checks, and updates.

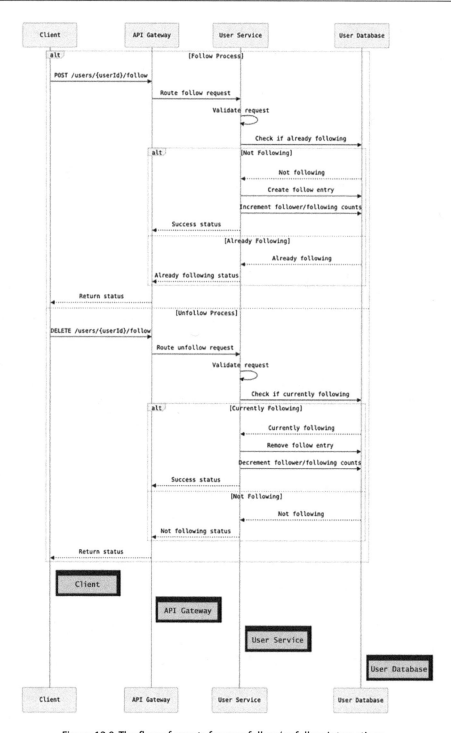

Figure 12.9: The flow of events for user follow/unfollow interactions

When a user follows or unfollows another user, the following takes place:

1. The **client** sends a POST or DELETE request to the /users/{userId}/follow endpoint with the user ID of the user to follow/unfollow.

2. The **User Service** validates the request and checks whether the user is authorized to perform the action.

3. **For a follow request**: The **User Service** creates a new entry in the Follow table, establishing the follower-followee relationship. The follower and following counts of both users are incremented.

4. **For an unfollow request**: The User Service removes the corresponding entry from the Follow table. The follower and following counts of both users are decremented.

5. The **User Service** returns a success status to the client.

We have explored the different flows of user interaction in our service, above. Now, let us understand the different modules that can help our User Service become performant and more useful:

- **Caching and optimization**: To improve the performance of the User Service and reduce the load on the database, user profile data is cached using a distributed caching system (e.g., Redis). Frequently accessed user profiles, such as those of popular users or users with a high number of followers, are cached to reduce the number of database queries. The cache is updated whenever a user's profile is modified or when the follower/following counts change.

- **Notifications and feed updates**: The User Service integrates with the Notification service and News Feed Service to handle relevant updates. When a user is followed by another user, the User Service sends a message to the Notification service to generate a notification for the followed user. The User Service also publishes an event to the News Feed Service to update the follower's news feed with the newly followed user's photos.

- **Security and privacy**: The User Service implements security and privacy measures to protect user data. User passwords are hashed securely using a strong hashing algorithm (e.g., bcrypt) before storing them in the database. Authentication tokens are generated using secure methods (e.g., JWT) and have an expiration time to prevent unauthorized access. User data is encrypted at rest and in transit to ensure confidentiality. Access controls are enforced to ensure that users can only modify their own profiles and follow/unfollow other users within the defined permissions.

By following this low-level design, the User Service provides a secure and efficient way to manage user profiles, authentication, and social interactions within the Instagram-like service. It leverages caching, event-driven architecture, and security best practices to ensure a seamless user experience.

We will now discuss some additional considerations to further enhance our service.

Additional considerations

When designing and implementing the Instagram-like service, there are several additional considerations and best practices to keep in mind. These considerations ensure the system is scalable, maintainable, and aligned with business requirements.

Let's explore some of these key aspects.

- **Hashtag functionality**:

  - Implement a hashtag system to allow users to categorize and discover photos based on specific topics or themes

  - Store hashtags as separate entities in the database, with a many-to-many relationship with photos

  - Provide API endpoints to search for photos based on hashtags and retrieve trending hashtags

  - Update the Photo Upload Service to extract hashtags from photo captions and associate them with the corresponding photos

- **Explore and discover features**:

  - Implement explore and discover features to help users find new and interesting content

  - Develop algorithms to suggest personalized recommendations based on user preferences, interests, and engagement history

  - Create curated collections or featured content showcasing popular or trending photos

  - Provide a search functionality to allow users to find photos, users, and hashtags based on keywords

- **Spam and abuse prevention**:

  - Implement measures to detect and prevent spam, abuse, and inappropriate content

  - Develop algorithms to identify and flag suspicious activities, such as automated likes, comments, or follow/unfollow patterns

  - Employ content moderation techniques, such as machine learning-based image classification, to detect and remove offensive or explicit content

  - Provide users with options to report and block abusive accounts or content

- **Internationalization and localization**:

  - Design the system to support internationalization and localization

  - Allow users to select their preferred language and locale settings

- Store user-generated content, such as captions and comments, in the original language and provide translation options if required

- **Monetization and business model**:

  - Develop a monetization strategy and business model for the Instagram-like service

  - Implement features for sponsored content, promoted posts, or advertising to generate revenue

  - Provide tools for businesses to create and manage their profiles, track metrics, and engage with their audience

  - Explore partnerships and integrations with e-commerce platforms to enable product tagging and in-app purchases

- **Monitoring, metrics, and observability**:

  - Implement a comprehensive logging system using a centralized log aggregator (e.g., ELK stack or Splunk) to collect and analyze logs from all microservices and components

  - Set up a metrics collection and visualization system (e.g., Prometheus and Grafana) to track **key performance indicators (KPIs)**, system health, and business metrics across the entire Instagram-like service

  - Utilize distributed tracing (e.g., Jaeger or Zipkin) to monitor and analyze request flows across microservices, helping to identify performance bottlenecks and optimize system interactions

  - Establish an alerting system with defined thresholds for critical metrics, and implement automated notifications (e.g., PagerDuty or OpsGenie) to ensure timely responses to potential issues or anomalies in the system

By considering these additional aspects and best practices, the Instagram-like service can be designed and implemented to be feature-rich, user-friendly, and aligned with business goals. It ensures a scalable and engaging platform that meets the evolving needs of users and stakeholders.

Summary

In this chapter, we have explored the system design of an Instagram-like service, covering various aspects such as functional and non-functional requirements, data modeling, scalability considerations, and architectural components. We have delved into the high-level design, laying out the overall architecture and the interaction between different services and components.

Furthermore, we have examined the low-level design of key services, including the Photo Upload Service, News Feed Service, and User Service. Each service has been designed to handle specific functionalities and work together to deliver a seamless user experience. We have discussed data storage, caching mechanisms, real-time updates, and performance optimization techniques to ensure scalability and efficiency.

Throughout the chapter, we have emphasized the importance of scalability, performance, and user engagement. The system has been designed to handle a large number of users, photos, and interactions by leveraging horizontal scaling, data partitioning, and distributed processing. Caching and content delivery networks have been employed to improve response times and reduce latency.

Some potential future considerations and enhancements for the Instagram-like service could include the following:

- Implementing advanced features such as stories, live streaming, and augmented reality filters to enhance user engagement and creativity

- Exploring machine learning and computer vision techniques to improve content recommendations, facial recognition, and object detection

- Integrating with social commerce platforms to enable seamless product tagging, in-app purchases, and influencer marketing

We can thus say that designing an Instagram-like service requires careful consideration of various aspects, from functional requirements to scalability and performance. By following the principles and best practices outlined in this chapter, developers and system architects can create a robust and scalable platform that meets the needs of users and stands the test of time.

In the next chapter, we will look at the system design of another popular service: Google Docs.

13

Designing a Service Like Google Docs

In the era of digital collaboration, file-sharing platforms have transformed the way people work together, and Google Docs stands out as one of the most powerful and intuitive services. With millions of active users, Google Docs has revolutionized the way people create, edit, and share documents online.

Designing a service such as Google Docs presents a unique set of challenges and opportunities. It requires a robust and scalable architecture that can handle real-time collaboration and a vast amount of user-generated content while providing a seamless and efficient user experience. In this chapter, we will explore the system design of a Google Docs-like service, delving into the key components, design decisions, and best practices involved in building a scalable and efficient file-sharing platform. By the end of this chapter, you will have a comprehensive understanding of the system design principles and techniques involved in building a service such as Google Docs. Let us start with the functional requirements of a service such as Google Docs.

In this chapter, we will cover the following topics:

- Functional requirements
- Non-functional requirements
- Data model
- Scale calculations
- High-level design
- Low-level design
- Additional considerations and best practices

Let us start by exploring the functional and non-functional requirements of a file-sharing system such as Google Docs.

Functional requirements

Before diving into the system design, it is crucial to define the functional requirements that specify what the file-sharing service should be capable of doing. These requirements lay the foundation for the entire design process and ensure that the system meets the needs of its users. Let's explore the key functional requirements:

- The user registration and authentication requirements are as follows:

  - Users should be able to create new accounts by providing necessary information such as a username, email, and password

  - The system should securely store user credentials and authenticate users upon login

  - User sessions should be managed efficiently to allow seamless access to the service

- The document creation, editing, and deletion requirements can be summed up as follows:

  - Users should be able to create new documents within the file-sharing service

  - The system should provide a rich text editor with formatting options, allowing users to create and edit documents collaboratively

  - Users should have the ability to delete documents they own or have appropriate permissions for

- Real-time collaboration and synchronization is a further requirement:

  - Multiple users should be able to edit the same document simultaneously, with their changes synchronized in real time

  - The system should handle concurrent edits efficiently, ensuring data consistency and avoiding conflicts

  - Users should be able to see the presence of other collaborators and their current cursor positions within the document

- Sharing and access control should be implemented as follows:

  - Users should be able to share documents with other users or external parties by generating shareable links or granting specific access permissions

  - The system should support different access levels, such as view-only, comment-only, or edit permissions

  - Document owners should have the ability to manage collaborators, revoke access, and modify sharing settings

- The version history and revision management features should include the following:

  - The file-sharing service should automatically save document revisions and maintain a version history

  - Users should be able to view and restore previous versions of a document

  - The system should provide a clear audit trail of changes made to a document, including the timestamp and the user responsible for each revision

- Commenting and suggestion features should be implemented as follows:

  - Users should have the ability to add comments to specific parts of a document, facilitating discussions and feedback

  - The system should support suggesting edits or changes, allowing collaborators to propose modifications without directly editing the document

  - Comments and suggestions should be easily visible and manageable within the document interface

- Search and organization features should include the following:

  - The file-sharing service should provide a powerful search functionality, enabling users to search for documents based on titles, content, or metadata

  - Users should be able to organize their documents into folders or categories for better management and accessibility

These functional requirements provide a comprehensive overview of the core features that a file-sharing service such as Google Docs should offer. By fulfilling these requirements, the system will enable users to create, collaborate, and share documents seamlessly, enhancing productivity and facilitating effective teamwork.

In the next section, we will explore the non-functional requirements that ensure the service remains scalable, reliable, and performant while meeting the functional requirements discussed previously.

Non-functional requirements

While functional requirements define what the system should do, non-functional requirements specify how the system should perform and behave. These requirements are critical in ensuring that the file-sharing service remains scalable, available, and reliable under various conditions. Let's discuss the key non-functional requirements:

- Scalability should be accounted for as follows:

  - The system should be designed to handle a large number of users and documents, accommodating growth and peak usage

- Horizontal scalability should be achieved by adding more servers and distributing the load across them

- The architecture should allow for easy scaling of individual components, such as the Document Service or Collaboration Service, independently

- High availability and reliability should be secured as follows:

  - The file-sharing service should be highly available, ensuring minimal downtime and quick recovery from failures

  - Redundancy should be implemented at various levels, including server redundancy, database replication, and geo-redundancy

  - The system should be designed to handle server failures, network outages, and data center disasters without significant impacts on the user experience

- The service should have low latency:

  - The service should provide real-time collaboration and synchronization with minimal latency

  - Users should be able to see each other's changes instantly, enabling a seamless and interactive collaboration experience

  - Techniques such as caching, efficient data retrieval, and optimized communication protocols should be employed to reduce latency

- Data consistency and integrity should be secured as follows:

  - The system should ensure the consistency and integrity of document data across all replicas and collaborators

  - Changes made by multiple users should be properly synchronized and merged to maintain a consistent document state

  - Conflict resolution mechanisms should be in place to handle simultaneous edits and prevent data corruption

By addressing these non-functional requirements, the file-sharing service can ensure a reliable, scalable, and performant user experience. It is essential to consider these requirements throughout the design process and make architectural decisions that align with these goals.

In the next section, we will explore the data model that forms the foundation of the file-sharing service, defining the entities and relationships necessary to support the functional requirements.

Data model

The data model is a crucial component of the file-sharing service, as it defines the structure and relationships of the data entities involved. A well-designed data model ensures efficient storage, retrieval, and manipulation of data while supporting the functional requirements of the system. Let's dive into the key entities and their relationships:

- The `User` entity represents the users of the file-sharing service, storing their basic information such as username, email, password hash, account creation, and last login timestamps. Each user is uniquely identified by their `user_id`.

- The `Document` entity represents the documents created and shared within the service. It contains the document title, content, `owner_id` (referencing the `User` who created it), creation and update timestamps, and current version number.

- The `Revision` entity stores the revision history of documents. Each revision is associated with a specific document and the user who made the changes. It includes the revised content and the timestamp of the revision.

- The `CollaboratorPermission` entity defines the access permissions granted to users for a specific document. It associates a user with a document and specifies the permission level (e.g., view, edit, or comment) and the timestamp when the permission was granted.

- The `Comment` entity represents comments made by users on specific documents. It contains the comment content, the associated document and user, and the creation timestamp.

- The `Suggestion` entity represents suggested changes or edits to a document. It includes the suggested content, associated document and user, status of the suggestion (e.g., pending, accepted, or rejected), and creation timestamp.

- The `Folder` entity allows users to organize their documents into folders. It contains the folder name, associated user, and creation timestamp.

- The `FolderDocument` entity represents the relationship between folders and documents, indicating which documents are contained within each folder.

Figure 13.1 illustrates the comprehensive data model for a Google Docs-like service. It showcases key entities such as `User`, `Document`, `Revision`, and `Folder`, along with their attributes and the relationships between them, providing a clear overview of the system's data structure and interconnections.

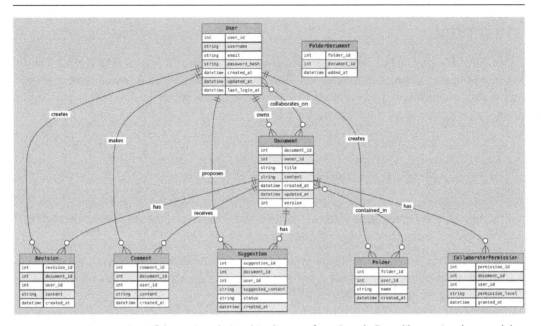

Figure 13.1: An overview of the entity relationship diagram for a Google Docs-like service data model

Let us talk about the different many-many, one-many, and many-one relationships in this data model next.

Relationships

There are several one-to-many and many-to-many relationships between the entities in this model. Let us highlight them.

The one-to-many relationships are as follows:

- **User to document**: A user can own multiple documents
- **Document to revision**: A document can have multiple revisions
- **Document to comment**: A document can have multiple comments
- **Document to suggestion**: A document can have multiple suggestions
- **User to folder**: A user can have multiple folders

The many-to-many relationships are as follows:

- **User to document** (through `CollaboratorPermission`): A user can collaborate on multiple documents, and a document can have multiple collaborators
- **Folder to document** (through `FolderDocument`): A folder can contain multiple documents, and a document can be present in multiple folders

By designing the data model with these entities and relationships, the file-sharing service can efficiently store and retrieve data related to users, documents, revisions, collaborations, comments, suggestions, and folder organization. The data model supports the functional requirements and enables the system to handle the complex interactions between users and documents.

In the next section, we will perform scale calculations to estimate the storage, bandwidth, and processing requirements of the file-sharing service based on the anticipated user base and usage patterns.

Scale calculations

To design a scalable file-sharing service, it is essential to estimate the storage, bandwidth, and processing requirements based on the expected user base and usage patterns. These calculations help in making informed decisions about the infrastructure and resources needed to support the service. Let's perform some scale calculations.

Assumptions

Here are some initial assumptions that we will use in our design and scale calculations:

- **Total number of users**: 10 million
- **Average number of documents per user**: 100
- **Average document size**: 50 KB
- **Average number of revisions per document**: 10
- **Average revision size**: 50 KB
- **Average number of collaborators per document**: 5
- **Average number of comments per document**: 20
- **Average comment size**: 1 KB
- **Average number of folders per user**: 10

Storage requirements

Here are the storage requirements for the system based on our previous assumptions:

- The document storage requirements are as follows:

  - **Total documents**: `10 million users × 100 documents/user = 1 billion documents`

  - **Total document storage**: `1 billion documents × 50 KB/document = 50 TB`

- The revision storage requirements are as follows:

  - **Total revisions**: `1 billion documents × 10 revisions/document = 10 billion revisions`

  - **Total revision storage**: `10 billion revisions × 50 KB/revision = 500 TB`

The comment storage requirements are as follows:

  - **Total comments**: `1 billion documents × 20 comments/document = 20 billion comments`

  - **Total comment storage**: `20 billion comments × 1 KB/comment = 20 TB`

- The user and metadata storage requirements are as follows:

  - **User metadata**: `10 million users × 1 KB/user = 10 GB`

  - **Folder metadata**: `10 million users × 10 folders/user × 1 KB/folder = 100 GB`

  - `CollaboratorPermission` **metadata**: 1 billion documents × 5 collaborators/document × 1 KB/permission = 5 TB

- The total storage requirements are as follows: `50 TB (documents) + 500 TB (revisions) + 20 TB (comments) + 5 TB (permissions) + 110 GB (user and folder metadata) ≈ 575 TB`

Bandwidth considerations

Based on our assumptions, here are the bandwidth needs for uploads and downloads:

- The document upload bandwidth needs are as follows:

  - **Average document uploads per day**: `10 million users × 1 document/user/ day × 50 KB/document = 500 GB/day`

- The document download bandwidth needs are as follows:

  - **Average document downloads per day**: `10 million users × 10 documents/ user/day × 50 KB/document = 5 TB/day`

- The collaboration and synchronization bandwidth needs are as follows:

  - **Average collaboration updates per day**: `10 million users × 10 documents/ user/day × 5 collaborators/document × 10 KB/update = 5 TB/day`

  - **Total daily Bandwidth**: `500 GB (uploads) + 5 TB (downloads) + 5 TB (collaboration) ≈ 10.5 TB/day`

Processing requirements

In our service, we need to process, render, format, and synchronize documents. Here are the estimates for these features for our service:

- The document rendering and formatting estimate is as follows:

  - **Peak document renders per second**: 10 million users × 10 documents/user/day ÷ 86400 seconds/day ≈ 1200 renders/second

- The revision comparison and merging estimate is as follows:

  - **Peak revision comparisons per second**: 10 million users × 10 documents/user/day × 1 revision/document/day ÷ 86400 seconds/day ≈ 1200 comparisons/second

- The real-time collaboration and synchronization estimates are as follows:

  - **Peak concurrent collaborators**: 10 million users × 10% concurrent usage × 5 collaborators/document ≈ 5 million concurrent collaborators

  - **Peak collaboration updates per second**: 5 million collaborators × 1 update/collaborator/minute ÷ 60 seconds/minute ≈ 83,000 updates/second

These calculations provide an estimate of the storage, bandwidth, and processing requirements for the file-sharing service based on the given assumptions. It's important to note that these numbers can vary based on actual usage patterns and the growth of the user base. The infrastructure should be designed to handle peak loads and should be easily scalable to accommodate future growth.

In the next section, we will propose a high-level design architecture that takes these scale requirements into account and leverages various building blocks to create a scalable and efficient system.

High-level design

Now that we have a clear understanding of the functional and non-functional requirements, as well as the scale calculations, let's dive into the high-level design of the file-sharing service. The goal is to create an architecture that is scalable, reliable, and efficient in handling the vast amount of documents, revisions, and user interactions. *Figure 13.2* shows the high-level design of the file-sharing system, which includes load balancers, API gateways, microservices for document management, collaboration, access control, caches, databases, and storage systems.

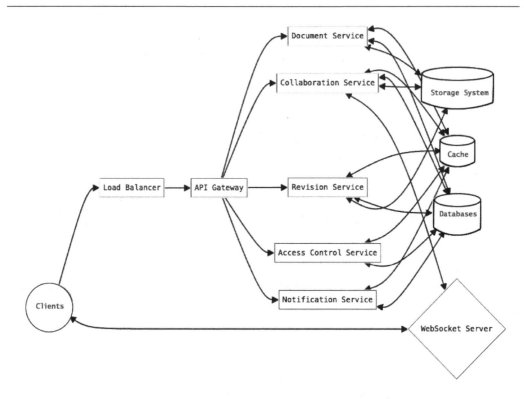

Figure 13.2: The high-level system design of a file-sharing service

Let us discuss the software components and modules shown in this *Figure 13.2*.

Software components and modules of high-level design

The following are the different components and modules:

- **Client-server architecture:**The file-sharing service will follow a client-server architecture, where clients (such as web browsers or mobile apps) communicate with the server-side components through APIs. The server-side components will handle the core functionality, data storage, and processing.

- **Load balancer:** To distribute the incoming traffic evenly across multiple servers, a load balancer will be placed in front of the server-side components. The load balancer will ensure that requests are efficiently routed to the appropriate servers based on factors such as server load, request type, and geographic location.

- **API Gateway:** An API Gateway will act as the entry point for all client requests. It will handle request routing, authentication, rate limiting, and request/response transformation. The API Gateway will expose well-defined APIs for various functionalities, such as document creation, editing, collaboration, and access control.

- **Microservices architecture**: To promote modularity, scalability, and maintainability, the server-side components will be designed as microservices. Each microservice will be responsible for a specific domain or functionality.

The main microservices in the file-sharing service will be as follows:

Document service:

- Handles document creation, retrieval, update, and deletion operations

- Manages document metadata and storage

- Communicates with the storage system to persist and retrieve document content

Collaboration service:

- Enables real-time collaboration and synchronization between multiple users

- Handles **Operational Transformation** (**OT**) and conflict resolution for concurrent editing

- Manages active user information and updates for collaborators

Revision service:

- Manages revision history and version control for documents

- Stores and retrieves document revisions

- Performs diff and merge operations for document revisions

Access control service:

- Handles user authentication and authorization

- Manages access permissions and sharing settings for documents

- Enforces access control policies for document operations

Notification service:

- Sends notifications and updates to users regarding document changes, comments, and collaboration events

- Integrates with email, push notifications, and real-time messaging systems

Caching:

- To improve performance and reduce the load on the backend services, a distributed caching layer (e.g., Redis) will be employed. The caching layer will store frequently accessed data, such as document metadata, user sessions, and commonly retrieved document content. Caching will help in serving data quickly and reducing the number of requests to the database and storage systems.

Databases:

- The file-sharing service will use a combination of databases to store structured and unstructured data:

Relational database (e.g., PostgreSQL):

- Stores metadata information such as user profiles, document metadata, access permissions, and collaboration details
- Provides ACID properties and supports complex querying and transactions

NoSQL database (e.g., MongoDB):

- Stores document revisions and content
- Provides scalability and flexibility for handling large amounts of unstructured data

Storage system:

- Document content, revisions, and attachments will be stored in a distributed storage system (e.g., Amazon S3 or Google Cloud Storage). Document types could be text files, videos, images, multimedia files, and so on. We may need to implement intelligent chunking or use multi-part APIs for some of these cloud storage services in order to support large sizes and parallelized uploads. The storage system will provide scalable and durable storage for large files, allowing for efficient retrieval and delivery to users.

Real-time communication:

- To enable real-time collaboration and synchronization, WebSocket connections can be established between the clients and the Collaboration Service. WebSocket allows for bidirectional communication, enabling instant updates and synchronization of document changes among collaborators.

Monitoring and logging:

- Comprehensive monitoring and logging mechanisms will be implemented to track the health and performance of the system. Metrics such as request latency, error rates, and resource utilization will be collected and visualized using tools such as Prometheus and Grafana. Centralized logging solutions (e.g., ELK stack) will be used to aggregate and analyze logs from all components.

Security and privacy:

- Security and privacy measures will be implemented throughout the system. User authentication and authorization will be handled using secure protocols such as OAuth. Sensitive data, such as passwords and access tokens, will be securely hashed and stored. Data encryption will be applied to protect user data both in transit and at rest. Regular security audits and penetration testing will be conducted to identify and address potential vulnerabilities.

This high-level design provides an overview of the key components and their interactions in the file-sharing service. It takes the scale requirements into account and employs various building blocks to create a scalable and efficient architecture.

In the subsequent sections, we will explore the low-level design of critical components, such as the Document Service, Collaboration Service, and Access Control Service, to gain a deeper understanding of their functionalities and interactions.

Low-level design

We will now cover the low-level design of several microservices that, together, form our service. Let us begin by evaluating the design of the document service.

Designing the document service

The Document Service is a critical component of the file-sharing system, responsible for handling document-related operations such as creation, retrieval, update, and deletion. It manages document metadata and interacts with the storage system to persist and retrieve document content. Let's dive into the low-level design of the Document Service. Let us first look at the API endpoints exposed by the service.

The Document Service exposes the following API endpoints:

- POST /documents: Create a new document

  - Request body: Document metadata (title, owner, etc.) and initial content

  - Response: Created document object with assigned document ID

- GET /documents/{documentId}: Retrieve a document by its ID

  - Response: Document object with metadata and content

- PUT /documents/{documentId}: Update a document's metadata or content

  - Request body: Updated document metadata and/or content

  - Response: Updated document object

- DELETE /documents/{documentId}: Delete a document by its ID

  - Response: Success or error message

- GET /documents/{documentId}/revisions: Retrieve the revision history of a document

  - Response: List of revision objects associated with the document.

Now, we will dive into the data models for this service.

Data models

Figure 13.3 illustrates the relationship between the `Document` and `Revision` entities, showing their attributes and the one-to-many relationship between them.

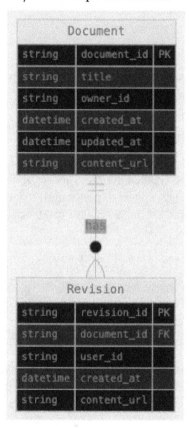

Figure 13.3: An overview of the entity relationship diagram for the Document Service

Now, let us look at the different workflows for creating, retrieving, updating, and deleting documents.

Document creation flow

Figure 13.4 shows the step-by-step process of creating a new document, including metadata storage and content upload.

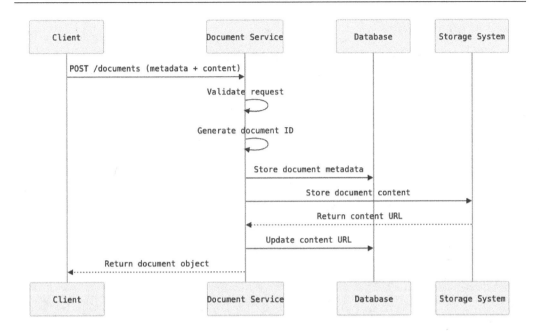

Figure 13.4: A document creation flow sequence

Here is the flow:

1. The **client** sends a POST request to the /documents endpoint with the document metadata and initial content.

2. The **Document Service** validates the request and generates a unique document ID.

3. The **Document Service** stores the document metadata in the relational database.

4. The initial content is stored in the storage system (e.g., S3), and the content URL is obtained.

5. The **Document Service** updates the document metadata with the content URL.

6. The created document object is returned to the client with the assigned document ID.

Let us now look at the retrieval flow.

Document retrieval flow

Figure 13.5 illustrates the process of retrieving a document, including fetching metadata from the database and content from the storage system.

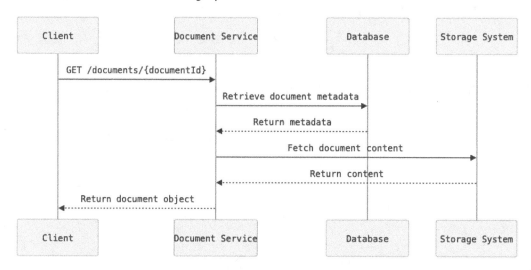

Figure 13.5: Document retrieval flow sequence diagram

Here is the flow:

1. The **client** sends a GET request to the `/documents/{documentId}` endpoint.

2. The **Document Service** retrieves the document metadata from the relational database based on the document ID.

3. If the document exists, the **Document Service** fetches the document content from the storage system using the content URL.

4. The document object, including metadata and content, is returned to the client.

Let us now look at the document update flow

Document update flow

Figure 13.6 illustrates the process of updating a document, showing how the system handles updates to metadata and/or content.

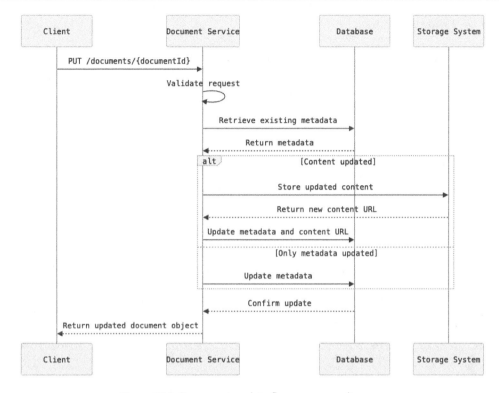

Figure 13.6: Document update flow sequence diagram

Here is the flow:

1. The **client** sends a PUT request to the /documents/{documentId} endpoint with the updated document metadata and/or content.

2. The **Document Service** validates the request and retrieves the existing document metadata from the relational database.

3. If the document exists, the **Document Service** updates the document metadata in the database.

4. If the content is provided in the request, the **Document Service** stores the updated content in the storage system and updates the content URL in the document metadata.

5. The updated document object is returned to the client.

Document deletion flow

Figure 13.7 shows the steps involved in deleting a document, including checks for document existence and removal from both database and storage.

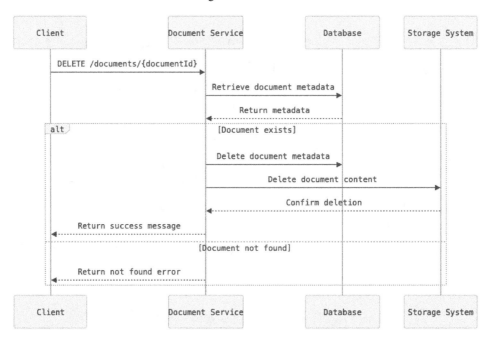

Figure 13.7: A document deletion flow sequence

Here is the flow:

1. The **client** sends a DELETE request to the /documents/{documentId} endpoint.

2. The **Document Service** retrieves the document metadata from the relational database based on the document ID.

3. If the document exists, the **Document Service** deletes the document metadata from the database.

4. The **Document Service** deletes the associated document content from the Storage System.

5. 5. A success message is returned to the client.

Let us now talk about how to get previous revisions of the content.

Revision history retrieval flow

Figure 13.8 illustrates the process of retrieving the revision history for a document, showing how the system fetches and returns the list of revisions.

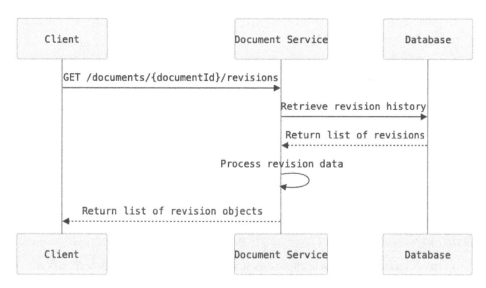

Figure 13.8: A revision history retrieval flow sequence

Here is the flow:

1. The **client** sends a GET request to the /documents/{documentId}/revisions endpoint.
2. The **Document Service** retrieves the revision history associated with the specified document ID from the relational database.
3. The list of revision objects, including metadata and content URLs, is returned to the client.

So far, we have discussed different document service flows. Let us now shift our attention to some performance and reliability considerations, specifically ones concerning caching and integrations.

The following are some of the performance and reliability considerations:

*   **Caching**: To improve the performance of document retrieval, the Document Service can utilize a distributed caching layer (e.g., Redis). Frequently accessed documents can be cached along with their metadata and content. When a document is requested, the Document Service first checks the cache. If the document is found in the cache, it is served directly from there. If the document is not found in the cache, the Document Service retrieves it from the database and storage system, stores it in the cache for future requests, and returns it to the client.

In addition to server-side caching, the system can leverage the user's browser cache to enhance performance and enable offline functionality. The client-side application can store local updates in the browser's cache (e.g., using `IndexedDB` or `LocalStorage`) and synchronize these updates with the server at regular intervals. This approach provides several benefits:

- **Improved responsiveness**: Users experience faster updates as changes are immediately reflected in the local cache

- **Offline support**: Users can continue working on documents even when they are disconnected from the internet

- **Resilience**: If the user experiences a brief network interruption, their work is not lost and can be synchronized once the connection is restored

- **Reduced server load**: By batching updates and synchronizing periodically, the system can reduce the number of real-time server requests

When implementing this strategy, it's important to handle conflict resolution for cases where multiple users may have made offline changes to the same document. The system should employ appropriate merging strategies or provide user interfaces for manual conflict resolution when necessary.

- **Error handling and consistency**: The Document Service implements error handling mechanisms to gracefully handle and propagate errors to the client. It includes appropriate error codes and messages in the API responses. The service also ensures data consistency by using transactions and atomic operations when modifying document metadata in the database. In case of failures during content storage or retrieval, the service retries the operations or returns appropriate error responses to the client.

- **Integration with other services**: The Document Service integrates with other services in the file-sharing system, such as the Collaboration Service and Access Control Service. When a document is created or updated, the Document Service notifies the Collaboration Service to handle real-time synchronization and presence updates. The Access Control Service is consulted to enforce access permissions and sharing settings for document operations.

- **Scalability and performance**: To ensure scalability and high performance, the Document Service can be deployed as a stateless microservice. Multiple instances of the service can run behind a load balancer to distribute the incoming requests. The relational database can be scaled horizontally using techniques such as sharding or partitioning based on document ID ranges. The storage system, such as S3, inherently provides scalability and durability for storing and retrieving document content.

By following this low-level design, the Document Service can efficiently handle document-related operations, ensure data consistency, and integrate with other services in the file-sharing system. The service is designed to be scalable, performant, and resilient to handle a large number of concurrent users and document interactions.

In the next section, we will explore the low-level design of the Collaboration Service, which enables real-time collaboration and synchronization between multiple users working on the same document.

Designing the Collaboration Service

The Collaboration Service is responsible for enabling real-time collaboration and synchronization between multiple users working on the same document. It handles OT, conflict resolution, and presence management to provide a seamless collaborative editing experience. *Figure 13.9* illustrates the key components of the Collaboration Service and its interactions with clients, other services, and infrastructure components.

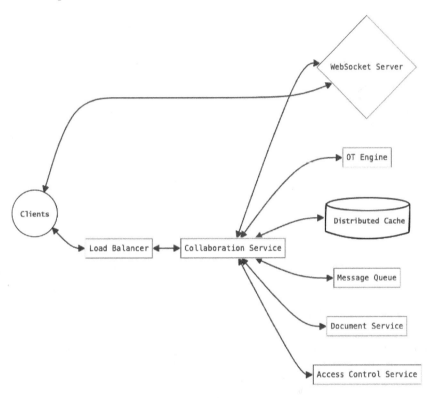

Figure 13.9: The Collaboration Service design

Let us now discuss the API endpoints exposed by the service and the flow.

The Collaboration Service exposes the following API endpoints:

- `POST /collaborate/{documentId}`: Join a collaborative editing session for a document

  - `Request body`: User ID and session information

- Response: Collaboration session details and initial document state

- WebSocket/collaborate/{documentId}: Establish a WebSocket connection for real-time collaboration

  - Clients send and receive collaboration-related events through the WebSocket connection

- POST /presence/{documentId}: Update user presence information for a document

  - Request body: User ID and presence status (e.g., online, offline, or idle)

  - Response: Success or error message

We will now discuss the collaboration flow with the help of a sequence diagram

Collaboration flow

Figure 13.10 shows the key steps in the collaboration process, including joining a session, real-time editing, and presence management.

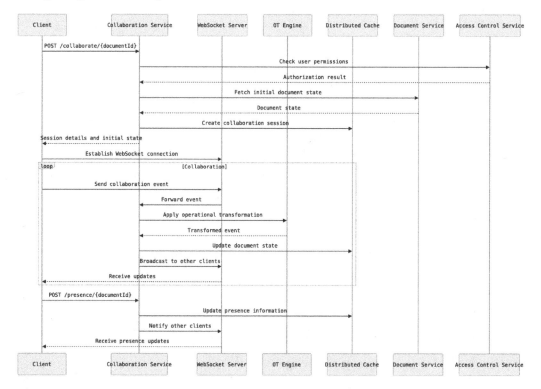

Figure 13.10: A collaboration flow sequence

Here are the steps:

1. When a user wants to collaborate on a document, the client sends a POST request to the / collaborate/{documentId} endpoint.

2. The **Collaboration Service** validates the request and checks the user's access permissions for the document.

3. If the user is authorized to collaborate, the **Collaboration Service** creates a new collaboration session and returns the session details and initial document state to the client.

4. The **client** establishes a WebSocket connection with the Collaboration Service using the / collaborate/{documentId} endpoint.

5. The **Collaboration Service** manages the WebSocket connections for all clients collaborating on the same document.

6. As users make changes to the document, the client sends collaboration events (e.g., insert, delete, or format) through the WebSocket connection.

7. The **Collaboration Service** receives the collaboration events and applies OT techniques to resolve conflicts and maintain a consistent document state across all collaborators.

8. The transformed events are broadcast to all connected clients through their respective WebSocket connections.

9. The clients receive the collaboration events and update their local document state accordingly, ensuring real-time synchronization.

We have explained the high-level collaboration workflow; let's dive into some nuances of collaboration, specifically those surrounding OT and presence management, next.

OT

OT is a technique used to achieve real-time collaboration while maintaining document consistency. The Collaboration Service implements OT algorithms to handle concurrent editing operations and resolve conflicts. The basic idea behind OT is to transform incoming operations based on the current document state and the history of previously applied operations. This ensures that the final document state is consistent across all collaborators, regardless of the order in which the operations were received.

The Collaboration Service maintains a server-side document state and applies incoming operations to it. It also keeps track of the operation history for each client. When a client sends a new operation, the Collaboration Service transforms the operation based on the client's operation history and the server-side document state. The transformed operation is then applied to the server-side document state and broadcast to all other clients.

Presence management

The Collaboration Service also manages user presence information to provide real-time awareness of collaborators' online status. When a user joins a collaboration session, the client sends a presence update to the `/presence/{documentId}` endpoint. The Collaboration Service updates the user's presence status and notifies other collaborators about the change.

The Collaboration Service maintains a presence map that associates each document with the list of collaborators and their presence status. It periodically checks for inactive or disconnected clients and updates their presence status accordingly. Clients can also explicitly notify the Collaboration Service when they go offline or become idle.

Scalability and performance

To ensure scalability and high performance, the Collaboration Service can be deployed as a stateless microservice. Multiple instances of the service can run behind a load balancer to distribute the incoming requests and WebSocket connections. The service can utilize a distributed caching layer (e.g., Redis) to store the server-side document state and presence information. This allows for quick access and synchronization across multiple service instances.

The Collaboration Service can also leverage message queues (e.g., Apache Kafka) to handle the broadcasting of collaboration events to connected clients. This decouples the event processing from the WebSocket connection handling and allows for asynchronous processing and scalability.

Integration with other services

The Collaboration Service integrates with the Document Service to retrieve and update document content. When a collaboration session is initiated, the Collaboration Service fetches the initial document state from the Document Service. As users make changes to the document, the Collaboration Service sends the updated content to the Document Service for persistence.

The Collaboration Service also interacts with the Access Control Service to enforce access permissions and ensure that only authorized users can collaborate on a document. It verifies the user's access rights before allowing them to join a collaboration session.

Error handling and resilience

The Collaboration Service implements error-handling mechanisms to handle and recover from various failure scenarios. It includes appropriate error logging and monitoring to detect and diagnose issues. In case of network disruptions or service failures, the Collaboration Service attempts to reconnect clients and synchronize the document state.

To ensure data integrity and consistency, the Collaboration Service implements transaction management and atomic operations when updating the server-side document state and presence information. It also includes data validation and sanitization to prevent malicious or invalid collaboration events from compromising the document state.

Security considerations

The Collaboration Service enforces security measures to protect the confidentiality and integrity of collaborative editing sessions. It uses secure communication protocols (e.g., HTTPS or WSS) to encrypt data in transit. Access control mechanisms are implemented to ensure that only authorized users can join collaboration sessions and access document content.

The Collaboration Service also implements rate limiting and throttling to prevent abuse and protect against denial-of-service attacks. It monitors and logs collaboration activities for auditing and security analysis purposes.

By following this low-level design, the Collaboration Service enables real-time collaboration and synchronization between multiple users, providing a seamless and interactive editing experience. The service is designed to handle concurrent editing operations, resolve conflicts, and maintain document consistency. It integrates with other services, such as the Document Service and Access Control Service, to provide a comprehensive collaborative editing solution.

In the next section, we will explore the low-level design of the Access Control Service, which handles user authentication, authorization, and access management for the file-sharing system.

Designing the Access Control Service

The Access Control Service is responsible for managing user authentication, authorization, and access permissions in the file-sharing system. It ensures that only authorized users can access and perform actions on documents based on their assigned roles and permissions. *Figure 13.11* illustrates the key components of the Access Control Service and its interactions with clients, databases, and other services in the system.

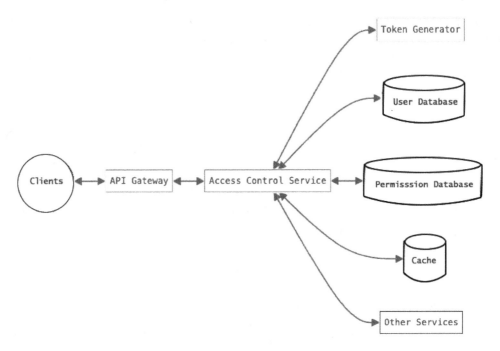

Figure 13.11: Access control service design

Let us discuss the API endpoints this service will expose next. The Access Control Service exposes the following API endpoints:

- POST /auth/login: Authenticate a user and generate an access token

  - Request body: User credentials (e.g., username and password)

  - Response: Access token and user information upon successful authentication

- POST /auth/logout: Invalidate a user's access token and log them out

  - Request body: Access token

  - Response: Success or error message

- GET /permissions/{documentId}: Get the access permissions for a document

  - Request parameters: Document ID

  - Response: List of user permissions for the document

- `POST /permissions/{documentId}`: Grant or update access permissions for a document

  - `Request body`: User ID, document ID, and permission level (e.g., read, write, or owner)

  - `Response`: Success or error message

- `DELETE/permissions/{documentId}/{userId}`: Revoke access permissions for a user in a document

  - `Request parameters`: Document ID and user ID

  - `Response`: Success or error message

Now that we have discussed the API endpoints, let us understand different authentication and authorization flows

Authentication flow

Figure 13.12 shows the steps involved in user authentication, from credential submission to access token generation

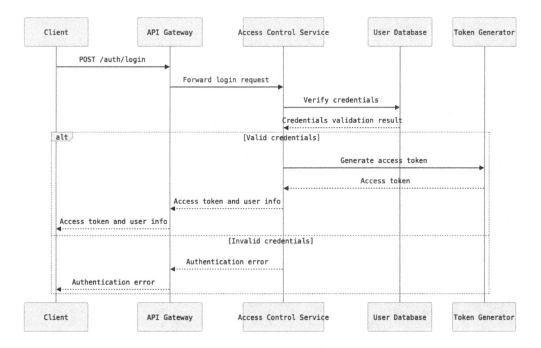

Figure 13.12: An authentication flow sequence

Here is the flow:

1. When a user logs in, the client sends a POST request to the /auth/login endpoint with the user's credentials.

2. The **Access Control Service** verifies the credentials against the user database.

3. If the credentials are valid, the **Access Control Service** generates an access token (e.g., **JSON Web Token (JWT)**) that contains the user's ID, role, and other relevant information.

4. The access token is returned to the client along with the user's information.

5. The client includes the access token in the headers of subsequent requests to authenticate and authorize the user.

Authorization flow

Figure 13.13 illustrates the process of authorizing a user's request, including token verification and permission checking.

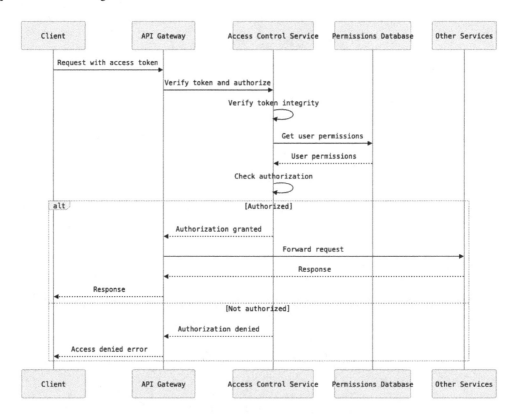

Figure 13.13: An authorization flow sequence

Here is the flow:

1. When a **client** makes a request to access a document or performs an action, it includes the access token in the request headers.

2. The **Access Control Service** intercepts the request and verifies the validity and integrity of the access token.

3. If the access token is valid, the **Access Control Service** extracts the user's ID and role from the token.

4. The **Access Control Service** retrieves the access permissions for the user on the requested document from the permissions database.

5. Based on the user's role and permissions, the **Access Control Service** determines whether the user is authorized to perform the requested action.

6. If the user is authorized, the request is forwarded to the appropriate service (e.g., Document Service) for further processing.

7. If the user is not authorized, the **Access Control Service** returns an appropriate error response.

Permission management

Figure 13.14 shows the steps involved in granting or updating permissions for a document, including the validation of the requester's permissions.

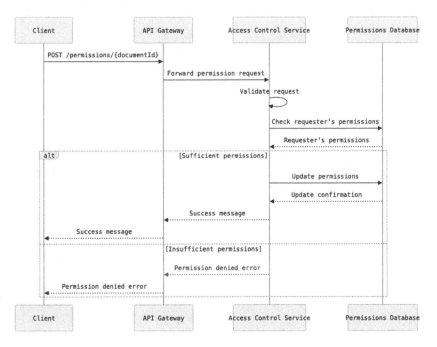

Figure 13.14: A permission management sequence

Here is the flow:

1. Document owners or users with appropriate permissions can grant or update access permissions for other users on a document.

2. The **client** sends a `POST` request to the `/permissions/{documentId}` endpoint with the user ID, document ID, and permission level.

3. The **Access Control Service** validates the request and verifies that the requesting user has sufficient permissions to grant or update permissions.

4. If the request is valid, the **Access Control Service** updates the permissions database with the new or updated permission entry.

5. The **Access Control Service** returns a success message to the client.

Revoking permissions

Figure 13.15 shows the steps involved in revoking a user's permissions for a document, including the validation of the requester's permissions and the removal of the permission entry from the database.

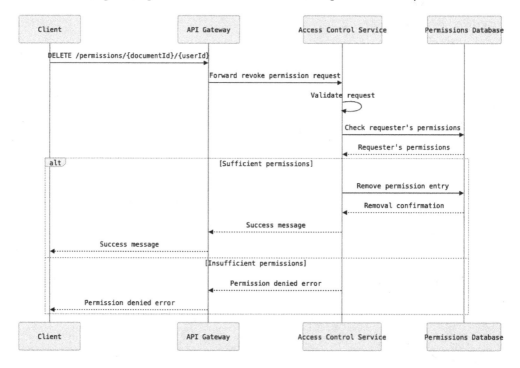

Figure 13.15: A revoking permissions sequence

Here is the flow:

1. Document owners or users with appropriate permissions can revoke access permissions for a user on a document.

2. The **client** sends a DELETE request to the /permissions/{documentId}/{userId} endpoint with the document ID and user ID.

3. The **Access Control Service** validates the request and verifies that the requesting user has sufficient permissions to revoke permissions.

4. If the request is valid, the **Access Control Service** removes the permission entry from the permissions database.

5. The **Access Control Service** returns a success message to the client.

Database design

The Access Control Service uses a database to store user information, access tokens, and permissions.

Figure 13.16 illustrates the structure of the database tables used by the Access Control Service, showing the relationships between Users, AccessTokens, Permissions, and Documents.

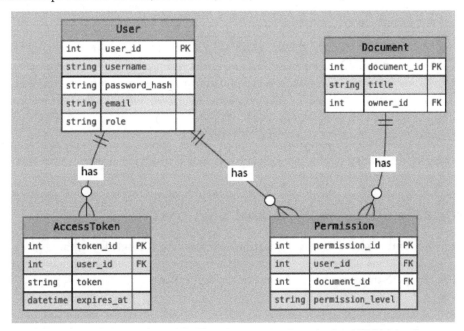

Figure 13.16: An entity-relationship diagram for an Access Control Service database

Some key points from the entity-relationship diagram are as follows:

- A User can have multiple `AccessTokens` and `Permissions`

- A `Document` can have multiple Permissions associated with it

- The Permission table acts as a junction table, connecting Users and Documents with specific permission levels

- The AccessToken table allows for the management of active sessions and token expiration

We will now look at how caching can help improve the performance of this service.

Caching and performance

To improve the performance of the Access Control Service, it can utilize a caching layer (e.g., Redis) to store frequently accessed permissions and user information. When a request is made to check permissions, the Access Control Service first checks the cache. If the permissions are found in the cache, they are served directly from there, reducing the load on the database.

The Access Control Service can also implement rate limiting and throttling mechanisms to prevent abuse and protect against unauthorized access attempts. It can monitor and log access control activities for auditing and security analysis purposes.

By following this low-level design, the Access Control Service provides a robust and secure mechanism for managing user authentication, authorization, and access permissions in the file-sharing system. It ensures that only authorized users can access and perform actions on documents based on their assigned roles and permissions. The service integrates with other components of the system to enforce access control throughout the application.

In the next section, we will discuss additional considerations and best practices for designing and implementing the file-sharing system, taking into account factors such as performance optimization, data consistency, and fault tolerance.

Additional considerations and best practices

When designing and implementing a file-sharing system such as Google Docs, there are several additional considerations and best practices to keep in mind. These considerations ensure that the system is robust, performant, and provides a seamless user experience. Let's explore some of these key aspects:

- **Performance optimization**:

  - Implement caching mechanisms at various levels of the system to reduce latency and improve response times. This can include caching frequently accessed documents, user permissions, and metadata.

- Utilize **Content Delivery Networks** (**CDNs**) to serve static assets, such as images and client-side scripts, from geographically distributed servers closer to the users.

- Optimize database queries and indexes to minimize the response time for common operations, such as document retrieval and permission checks.

- Employ lazy loading techniques to load document content and collaborators incrementally as needed, reducing the initial load time.

- Implement pagination and limit the number of results returned by API endpoints to avoid overloading the system and improve performance.

- **Data consistency and concurrency**:

  - Ensure that the system maintains data consistency across all services and components, especially in scenarios involving concurrent editing and real-time collaboration.

  - Implement optimistic concurrency control mechanisms, such as version numbers or timestamps, to detect and resolve conflicts when multiple users are editing the same document simultaneously.

  - Use distributed locking or synchronization techniques to prevent data inconsistencies and ensure atomic operations when necessary.

  - Employ eventual consistency models where appropriate, allowing for temporary inconsistencies in favor of higher availability and performance.

- **Fault tolerance and resilience**:

  - Design the system to be resilient to failures and to be able to recover gracefully from errors and exceptions.

  - Implement error handling and logging mechanisms to capture and diagnose issues promptly.

  - Use circuit breakers and retry mechanisms to handle temporary failures and prevent cascading failures across services.

  - Employ redundancy and replication techniques to ensure high availability and minimize the impact of hardware or network failures.

  - Regularly backup data and implement disaster recovery procedures to protect against data loss and ensure business continuity.

- **Scalability and elasticity**:

  - Design the system to be horizontally scalable, allowing for the addition of more instances of services as the user base and traffic grow.

  - Utilize auto-scaling techniques to automatically adjust the number of service instances based on the incoming load, ensuring optimal resource utilization and cost efficiency.

- Implement load balancing mechanisms to evenly distribute the workload across multiple instances of services.

- Use message queues and asynchronous processing to decouple services and handle spikes in traffic or resource-intensive tasks.

- **Continuous Integration and Continuous Deployment (CI/CD):**

- Implement a robust CI/CD pipeline to automate the building, testing, and deployment processes for the file-sharing system.

- Use version control systems (e.g., Git) to manage the codebase and enable collaborative development.

- Automate unit tests, integration tests, and end-to-end tests to catch bugs and ensure the stability and reliability of the system.

- Utilize containerization technologies (e.g., Docker) and orchestration platforms (e.g., Kubernetes) to streamline the deployment and management of services.

- Implement blue-green deployments, canary releases, or rolling updates to minimize downtime and risk during deployments.

By considering these additional aspects and best practices, the file-sharing system can be designed and implemented to be performant, scalable, secure, and user-friendly. It is important to continuously monitor, measure, and iterate on the system based on user feedback, performance metrics, and changing requirements to ensure its long-term success and adoption.

Summary

In this chapter, we explored the system design of a file-sharing service such as Google Docs, covering various aspects such as functional and non-functional requirements, data modeling, scalability considerations, and architectural components. We delved into the high-level design, laying out the overall architecture and the interaction between different services and components.

Throughout the chapter, we emphasized the importance of scalability, performance, and user experience. The system has been designed to handle a large number of users, documents, and concurrent editing operations by leveraging distributed architectures, caching mechanisms, and real-time collaboration techniques. We learned that the use of microservices architecture allows for the independent scaling and deployment of individual components, ensuring flexibility and maintainability.

We also discussed the critical role of data consistency and integrity in a file-sharing system. The use of OT algorithms and conflict resolution techniques ensures that documents remain consistent across multiple collaborators and editing sessions. The data model has been designed to efficiently store and retrieve document metadata, revisions, and permissions, enabling fast access and modification.

Furthermore, we discussed additional considerations and best practices that contribute to the success and user adoption of a file-sharing system. We learned that performance optimization techniques, such as caching and lazy loading, help in delivering a responsive and smooth user experience. Collaboration features, such as real-time editing, comments, and suggestions, foster teamwork and productivity, as we learned in this chapter. We also learned that CI/CD practices enable rapid iteration and delivery of new features and bug fixes. In the next chapter, we will look into designing a service such as Netflix.

14

Designing a Service Like Netflix

In today's digital age, video streaming services have revolutionized the way we consume entertainment. Netflix, one of the pioneers in this industry, has set a high bar for delivering a seamless and personalized video streaming experience to millions of users worldwide. Designing a service such as Netflix requires careful consideration of various aspects, including scalability, performance, user experience, and reliability.

In this chapter, we will delve into the system design principles and components necessary to build a robust and scalable video streaming service such as Netflix. To understand how to design a service such as Netflix, we will cover the following topics:

- Functional requirements

- Non-functional requirements

- Designing the data model

- Scale calculations

- High-level design

- Low-level design

By the end of this chapter, you will have a comprehensive understanding of the building blocks and design decisions involved in creating a service such as Netflix. Let's begin by examining the functional requirements that define the core features and capabilities of a Netflix-like service.

Functional requirements

Before we embark on designing our Netflix-like service, it's crucial to define the functional requirements that specify what our system should be capable of doing. These requirements will guide our design decisions and ensure that we build a platform that meets the needs and expectations of our users.

Let's outline the key functional requirements:

- **User registration and authentication**:

  - Users should be able to create new accounts by providing necessary information, such as email, password, and profile details

  - The system should securely store user credentials and authenticate users upon login

  - User sessions should be managed efficiently to allow seamless access to the service across different devices

- **Content browsing and search**:

  - Users should be able to browse and explore a vast catalog of movies, TV shows, and other video content

  - The system should provide a user-friendly interface to navigate through different categories, genres, and recommendations

  - Users should be able to search for specific titles, actors, directors, or keywords to find desired content quickly

- **Video playback and streaming**:

  - Users should be able to play videos seamlessly, with minimal buffering and high-quality streaming

  - The system should support various video resolutions and adapt to the user's network bandwidth for an optimal playback experience

  - Users should have control over video playback, including play, pause, seek, and resume functionalities

- **User profiles and preferences**:

  - Each user should have a personalized profile that captures their viewing preferences, watch history, and ratings

  - There should be multiple profiles supported per account so that each profile can be personalized separately

  - Users should be able to update their profile information, manage their viewing preferences, and set parental controls

- **Recommendations and personalization**:

  - Based on a user's preferences, their history of watched content, and the viewing history of similar users, the system should provide content recommendations

  - Recommendations should be generated using sophisticated algorithms that analyze user data and identify similar content

  - The system should continuously learn and adapt to user feedback to refine and improve the accuracy of recommendations over time

- **Watchlist and viewing history**:

  - Users should be able to add titles to their watchlist for future viewing and easily access their viewing history

  - The system should sync the watchlist and viewing history across different devices, allowing users to seamlessly continue watching from where they left off

- **Offline viewing**:

  - The system should support offline viewing functionality, allowing users to download selected titles and watch them without an internet connection.

  - Downloaded content should be securely stored on the user's device and have an expiration mechanism to comply with licensing agreements.

These functional requirements lay the foundation for our Netflix-like service, defining the core features and capabilities that our system must deliver to provide an immersive and personalized video streaming experience to our users.

In the next section, we'll explore the non-functional requirements that ensure our service remains scalable, reliable, and performant while handling a large user base and streaming massive amounts of video content.

Non-functional requirements

While functional requirements define what our Netflix-like service should do, non-functional requirements specify how the system should perform and behave under various conditions. These requirements are crucial for ensuring a smooth, reliable, and efficient streaming experience for our users. Let's discuss the key non-functional requirements:

- **Scalability and performance**:

  - The system should be designed to handle a large number of concurrent users and a growing library of video content

- It should be able to scale horizontally by adding more servers to handle increased traffic and storage needs

- The architecture should allow the efficient distribution of content across multiple servers and data centers to ensure fast and reliable streaming

- Caching mechanisms should be employed at various levels to reduce latency and improve performance

- **Availability and reliability**:

  - The system should be highly available, ensuring minimal downtime and quick recovery from failures

  - It should be designed with redundancy and fault tolerance in mind, using techniques such as data replication, load balancing, and automatic failover

  - The system should be able to handle server failures, network outages, and data center disasters without significant impact on user experience

  - Regular data backups and disaster recovery mechanisms should be in place to prevent data loss and ensure business continuity

- **Content delivery and streaming quality**:

  - The system should leverage a **Content Delivery Network** (**CDN**) to efficiently deliver video content to users across different geographical locations

  - Depending on the user's device and its location, our system should dynamically adjust video quality by supporting adaptive bitrate streaming

  - The streaming infrastructure should be optimized to minimize buffering, reduce latency, and provide a seamless playback experience

  - The system should handle a large number of concurrent streams and support various video formats and codecs

By addressing these non-functional requirements, our Netflix-like service will be able to deliver a high-quality, reliable, and scalable streaming experience to our users. These requirements will shape our architectural decisions and guide us in designing a robust and efficient system.

Let's now examine the data model that will serve as the backbone of our service, defining the entities and relationships necessary to support the functionality of our Netflix-like platform.

Designing the data model

Designing an efficient and scalable data model is essential for our Netflix-like service. The data model defines the structure and relationships of the entities involved in the system, ensuring that data is stored and retrieved effectively. Let's explore the key entities and their relationships:

- **User**: The central entity, representing a customer, which can have multiple profiles associated with it, allowing for personalized experiences within a single account

- **Profile**: Linked to a user, this entity represents individual viewing preferences and settings and is associated with the `WatchHistory`, `Watchlist`, and `Rating` entities to track personalized interactions with content

- **Movie**: A standalone content entity that has its own metadata, and it can be part of a user's `WatchHistory` and `Watchlist` and receive `Ratings` from `Profiles`

- **TVShow**: Represents a series that contains multiple `Episodes`, has its own metadata, and, like `Movies`, can be part of `WatchHistory`, `Watchlist`, and receive `Ratings`

- **Episode**: Belongs to a `TVShow` and has its own metadata, allowing for granular tracking of viewing progress within a series through the WatchHistory entity

- **WatchHistory**: Tracks the viewing activity of a `Profile`, linking to either a `Movie`, `TVShow`, or specific `Episode`, enabling features such as resuming playback and viewing statistics

- **Watchlist**: Allows `Profiles` to save `Movies` or `TVShows` for future viewing, supporting content discovery and personalized recommendations

- **Rating**: Enables `Profiles` to provide feedback on `Movies` or `TVShows`, which can be used for personalized recommendations and overall content quality assessment

- **ContentMetadata**: Stores technical details about the content (`Movie`, `TVShow`, or `Episode`), such as video quality and file size, supporting adaptive streaming and content delivery optimization

These entities and their relationships form a comprehensive data model that supports the core functionalities of a Netflix-like streaming service, enabling personalized user experiences, content management, and interaction tracking.

Figure 14.1's entity relationship diagram illustrates the comprehensive data model for a Netflix-like streaming service, showcasing the key entities such as `User`, `Profile`, `Movie`, `TVShow`, `Episode`, `WatchHistory`, `Watchlist`, `Rating`, and `ContentMetadata`, along with their attributes and the relationships between them.

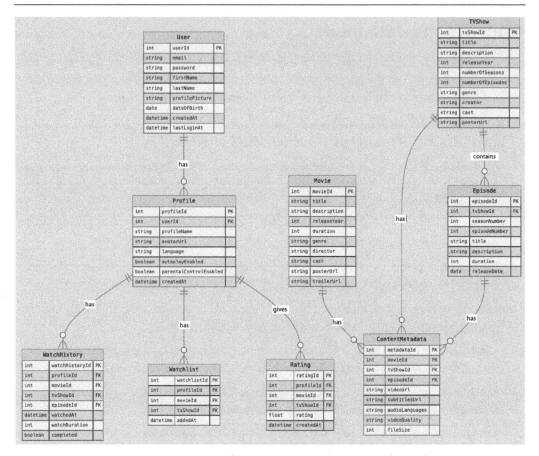

Figure 14.1: An overview of the entity relationship diagram for Netflix

Let's now discuss the many-many, one-many, and many-one relationships between the different entities in this data model design.

Relationships

Lets us highlight the different one-to-many relationships in our data model.

- OnetoMany:

  - `User` to `Profile`: A user can have multiple profiles

  - `Movie` to `ContentMetadata`: A movie can have multiple versions of metadata (e.g., different video qualities or languages)

- `TVShow` to `Episode`: A TV show can have multiple episodes

- `TVShow` to `ContentMetadata`: A TV show can have multiple versions of metadata

- `Episode` to `ContentMetadata`: An episode can have multiple versions of metadata

- `Profile` to `WatchHistory`: A profile can have multiple watch history entries

- `Profile` to `Watchlist`: A profile can have multiple watchlist entries

- `Profile` to `Rating`: A profile can have multiple ratings

The preceding data model captures the essential entities and relationships required for our Netflix-like service. It allows us to store and manage user information, profiles, movies, TV shows, episodes, watch history, watchlists, ratings, and content metadata.

By designing the data model in a structured and efficient manner, we can ensure that our system can handle the storage and retrieval of large amounts of data while supporting the key functionalities of our streaming platform.

In the next section, we'll perform capacity estimation and scaling calculations to determine the storage, bandwidth, and processing requirements of our Netflix-like service, based on the expected user base and usage patterns.

Scale calculations

To design a scalable Netflix-like service, it's crucial to estimate the storage, bandwidth, and processing requirements, based on the expected user base and usage patterns. These estimations will help us determine the necessary infrastructure and resources needed to support the service. Let's perform some capacity estimation and scaling calculations by making some basic assumptions.

Here are some of the assumptions we will make while designing our Netflix-like service.

Assumptions

- **The total number of users**: 50 million

- **The average number of active users per day**: 10 million

- **The average number of videos watched per user per day**: 3

- **The average video duration**: 1.5 hours

- **The average video file size (HD quality)**: 3 GB

Next, based on these assumptions, let's estimate the storage for our service

Storage estimation

Storage will typically include both data and metadata storage. In our case, we have videos, thumbnails, users, and so on, so let's try to estimate, based on the preceding assumptions, how much storage we will need to make our service functional:

- **Video storage:**

  - **The total number of movies:** `10,000`

  - **The total number of TV shows:** `5,000`

  - **The average number of episodes per TV show:** `30`

  - **The total number of episodes:** `5,000 × 30 = 150,000`

  - **The total video files:** `10,000 (movies) + 150,000 (episodes) = 160,000`

  - **The total video storage:** `160,000 × 3 GB = 480 PB`

- **User and metadata storage:**

  - **User data:** `50 million × 1 KB = 50 GB`

  - **Profile data:** `50 million × 5 profiles × 1 KB = 250 GB`

  - **Watch history and ratings data:** `50 million × 5 profiles × 1 MB = 250 TB`

  - **Content metadata:** `160,000 × 10 KB = 1.6 GB`

  - **Total storage:** `480 PB (video) + 250 TB (user and metadata) ≈ 480 PB`

We now have a good estimate of the storage needs, so we will move on to bandwidth estimations.

Bandwidth estimation

One of the ways Netflix stands out from the competition is its seamless user experience, made possible due to accurate bandwidth estimation and management. We will attempt to estimate our bandwidth needs to support our users.

- **The daily video streaming bandwidth:**

  - **The average video file size:** `3 GB`

  - **The average number of videos streamed per day:** `10 million users × 3 videos = 30 million`

  - **The daily streaming bandwidth:** `30 million × 3 GB = 90 PB/day`

- **Peak bandwidth requirements**:

  - **Peak concurrent users**: `1 million`

  - **Peak bandwidth**: `1 million × 3 GB / 1.5 hours = 2 TB/s`

Processing estimation

Netflix is run on different systems that range from smartphones and tablets to browsers and television sets. For each of these systems, the same movie or content is rendered, stored, and transferred differently to make the experience compelling. Therefore, we need to perform some pre- and post-processing of content, which is very typical for video streaming services such as Netflix. Let's estimate the scale for these:

- **Video encoding and transcoding**:

  - **The number of new videos added per day**: `100`

  - **The average video duration**: `1.5 hours`

  - **The encoding time**: `1.5 hours × 5 (multiple bitrates) = 7.5 hours per video`

  - **Daily encoding processing**: `100 videos × 7.5 hours = 750 hours`

- **Recommendation processing**:

  - **The number of active users per day**: `10 million`

  - **The recommendation requests per user per day**: `10`

  - **The total recommendation requests per day**: `10 million × 10 = 100 million`

  - **The processing time per request**: `100 ms`

  - **Daily recommendation processing**: `100 million × 100 ms = 2.7 hours`

So far, we have discussed different scale calculations for our service, but there are some inherent scaling considerations that we would like to explicitly spell out. The following are some of the scaling considerations:

- **Horizontal scaling**:

  - Use load balancers to distribute traffic across multiple application servers

  - Scale the number of application servers, based on the incoming traffic and processing requirements

  - Utilize auto-scaling techniques to dynamically adjust the number of servers, based on demand

- **Caching**:

    - Implement caching at various levels (e.g., CDN, application server, and database) to reduce the load on backend systems and improve performance

    - Cache frequently accessed data such as popular videos, user profiles, and recommendations

- **CDN**:

    - Utilize a CDN to distribute video content globally and reduce latency for users.

    - Store video files on CDN servers located closer to the users for faster streaming.

- **Database sharding and replication**:

    - Shard the database to distribute data across multiple servers and handle increased storage and traffic

    - Replicate data to ensure high availability and fault tolerance

These estimations provide a starting point for designing the infrastructure and scaling our Netflix-like service. It's important to note that these numbers can vary, based on actual usage patterns, video quality, and the growth of the user base. Regular monitoring, performance analysis, and capacity planning should be conducted to adapt to changing requirements.

In the next section, we'll propose a high-level architecture that takes into account these capacity estimations and incorporates the necessary components and services to build a scalable and efficient video streaming platform.

High-level design

Now that we have a clear understanding of the functional and non-functional requirements, data model, and capacity estimations, let's design a high-level architecture for our Netflix-like service. The architecture should be scalable, reliable, and efficient in handling the massive amount of video content and user traffic. *Figure 14.2* shows the high-level architecture of our service. Let's dive into the flow and understand this architecture in detail.

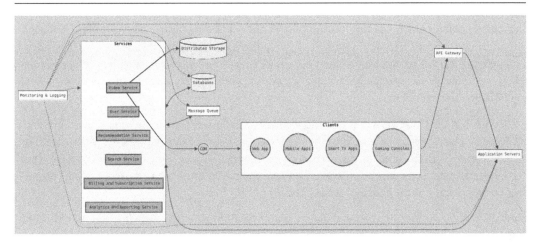

Figure 14.2: The Netflix architecture diagram

This diagram illustrates the comprehensive architecture of a Netflix-like streaming service, showcasing the flow from client applications through various components, including the API gateway, application servers, specialized services, CDN, storage systems, and supporting infrastructure such as a Message Queue and Monitoring and Logging.

At the heart of our architecture are the **client applications**. These include the web application, mobile applications (iOS and Android), smart TV applications, and gaming console applications. These client applications serve as the primary interface for users to interact with our streaming service.

When a user makes a request through one of the client applications, it first reaches the **API gateway**. The API gateway acts as the entry point for all client requests and plays a crucial role in the system. It handles request routing, authentication, and rate limiting, ensuring that only authorized requests are forwarded to the appropriate backend services. The API gateway also provides a unified API interface, making it easier for the front-end applications to communicate with the various backend components.

Behind the API gateway, we have the **application servers**. These servers handle the business logic and processing of client requests. They communicate with various backend services and databases to retrieve and manipulate data as needed. The application servers are designed to scale horizontally, based on traffic and processing requirements, ensuring that the system can handle a growing user base and increasing demands.

One of the critical components of our architecture is the **Video Service**. This service is responsible for handling video storage, encoding, and streaming. When a user uploads a video or when new content is added to the platform, the Video Service stores the video files in a distributed storage system. It then encodes the videos into multiple formats and bitrates to support adaptive streaming. Adaptive streaming allows the video quality to adjust dynamically, based on the user's network conditions, providing a smooth and uninterrupted viewing experience.

To efficiently deliver video content to users, the Video Service integrates with a CDN. The CDN distributes video content globally, bringing it closer to the users and reducing latency. It caches video files on servers located strategically around the world, allowing users to stream videos from a server that is geographically close to them. This significantly improves streaming performance and ensures a high-quality user experience.

Next, let's explore the **User Service**. This service manages user authentication, authorization, and profile data. When a user signs up or logs in to the platform, the User Service handles the authentication process. It securely stores user information, such as credentials and profile data, in a database. The User Service also manages user permissions and access controls, ensuring that users can only access the content and features they are authorized to use.

To provide personalized video recommendations to users, we have the **Recommendation Service**. This service is responsible for generating tailored suggestions based on a user's viewing history, preferences, and behavior. It analyzes various data points, such as the videos a user has watched, their ratings, and their interactions with the platform. By leveraging machine learning algorithms, the Recommendation Service can identify patterns and make accurate recommendations, enhancing the user experience and encouraging user engagement.

Another essential component of our architecture is the **Search Service**. With a vast library of video content, it's crucial to provide users with a powerful search functionality. The Search Service indexes video metadata, including titles, descriptions, genres, and tags, making it easy for users to find the content they are looking for. It supports features such as autocomplete and fuzzy matching, ensuring that users can quickly discover relevant videos even if they don't remember the exact title or spelling.

To handle user subscriptions, billing, and payments, we have the **Billing and Subscription Service**. This service integrates with payment gateways and manages recurring billing for users who have subscribed to our platform. It provides APIs for subscription management, allowing users to upgrade, downgrade, or cancel their subscriptions. The Billing and Subscription Service also generates invoices and handles payment processing securely.

Monitoring and analyzing user behavior and system performance is crucial for making data-driven decisions and ensuring a smooth user experience. The **Analytics and Reporting Service** takes care of collecting and analyzing various metrics and user behavior data. It generates insights and reports that help a business make informed decisions, such as identifying popular content, optimizing recommendations, and improving the overall service.

To support the communication and coordination between different services, we utilize a **Message Queue**. The Message Queue facilitates asynchronous communication and enables an event-driven architecture. It decouples services, allowing them to communicate and exchange data without direct dependencies. For example, when a new video is uploaded, the Video Service can send an encoding job to the Message Queue, which can then be picked up and processed by a separate encoding service. This ensures that tasks are processed efficiently and reliably, even under a high load.

Lastly, to ensure the reliability and stability of our system, we implement comprehensive **Monitoring and Logging**. We collect logs and metrics from all components of the architecture and centralize them for analysis and troubleshooting. This allows us to quickly identify and resolve any issues that may arise. We also set up dashboards and alerts to provide real-time visibility into the system's health and performance.

In summary, our high-level architecture is designed to handle the scale and complexity of a Netflix-like streaming service. The client applications interact with the API gateway, which routes requests to the appropriate backend services. The Video Service handles video storage, encoding, and streaming, while the User Service manages user authentication and profile data. The Recommendation Service provides personalized video suggestions, and the Search Service enables users to find content easily. The Billing and Subscription Service handles payments and subscriptions, and the Analytics and Reporting Service provides valuable insights. The Message Queue facilitates communication between services, and Monitoring and Logging ensure the system's reliability and performance.

In the next sections, we'll dive deeper into the design and implementation details of low-level system design and focus on some key components, such as the Video Service, User Service, Recommendation Service, and CDN.

Low-level system design

In this section, we will pick some of the main microservices in our system and try to delve into their APIs and design. Note that we are not going to cover each part of our system, just some of the main services to give you a feel for the design and implementation details.

Video Service

The Video Service is a critical component of our Netflix-like architecture, responsible for handling video storage, encoding, and streaming. Let's dive into the design and implementation details of the Video Service.

Figure 14.3 shows the high-level architecture of the service.

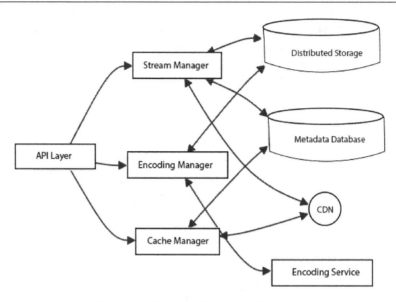

Figure 14.3: The Video Service architecture

Let's first discuss the API endpoints exposed by the service:

- POST /videos: Upload a new video file:

  - Request body: The video file and metadata (title, description, duration, etc.).

  - Response: The video ID and status.

- GET /videos/{videoId}: Retrieve video metadata:

  - Response: Video metadata (title, description, duration, etc.).

- GET /videos/{videoId}/stream: Stream the video content:

  - Request parameters: Video ID, quality/bitrate, offset.

  - Response: Video chunk or segment.

Now that we have discussed the API endpoints, let's look at the different workflows for this service. We will use sequence diagrams to explain these.

Video upload and storage

Figure 14.4 shows the video upload and storage workflow sequence diagram.

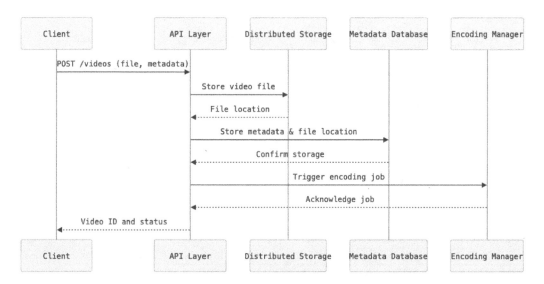

Figure 14.4: Video upload and storage

Here is the flow:

1. When a new video is uploaded, the **Video Service** receives the video file and metadata through the API endpoint.

2. The video file is stored in a distributed storage system, such as Amazon S3 or **Hadoop Distributed File System (HDFS)**, for scalable and durable storage.

3. The video metadata is stored in a database, such as PostgreSQL or Cassandra, along with a reference to the video file location in the distributed storage.

Let's now look at video encoding and transcoding workflows.

Video encoding and transcoding

Figure 14.5 shows the video encoding and transcoding sequence diagram.

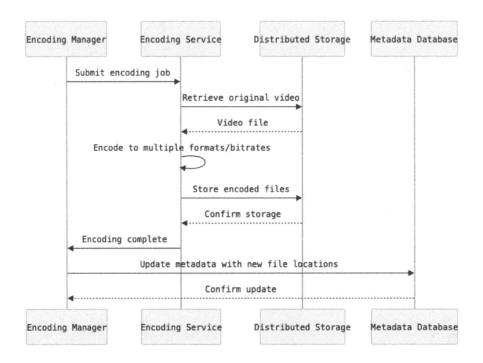

Figure 14.5: Video encoding and transcoding

Here is the flow:

1.   After a video is uploaded, the **Video Service** triggers a video encoding job to convert the video into multiple formats and bitrates.

2.   The encoding job can be handled by a separate encoding service or a cluster of encoding workers.

3.   The video is transcoded into different formats (e.g., MP4, HLS, and DASH) and bitrates to support adaptive streaming and various client devices.

4.   The encoded video files are stored in the distributed storage system, and their locations are updated in the video metadata database.

Next, let's look at the flow for the video streaming aspect of our service.

Video streaming

Figure 14.6 shows the video streaming sequence diagram.

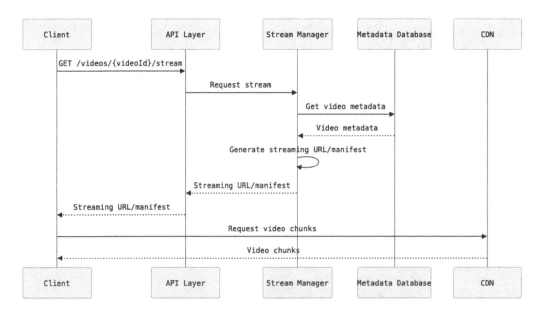

Figure 14.6: The video streaming sequence diagram

Here is the flow:

1. When a user requests to stream a video, the **client** application sends a request to the Video Service API endpoint.

2. The **Video Service** retrieves the video metadata from the database and determines the appropriate video format and bitrate, based on the client's capabilities and network conditions.

3. The **Video Service** generates a streaming URL or manifest file (e.g., an HLS playlist or DASH manifest) that contains the information needed by the client to stream the video.

4. The **client** application uses the streaming URL or manifest file to request video chunks or segments from the **CDN**, or directly from the **Video Service**.

5. The **CDN** or **Video Service** serves the video chunks or segments to the **client** application, which plays them in the correct order to provide a continuous streaming experience.

Next, let's look at the flow for content delivery and appropriate caching.

Caching and content delivery

Figure 14.7 shows the sequence diagram for content delivery and appropriate caching.

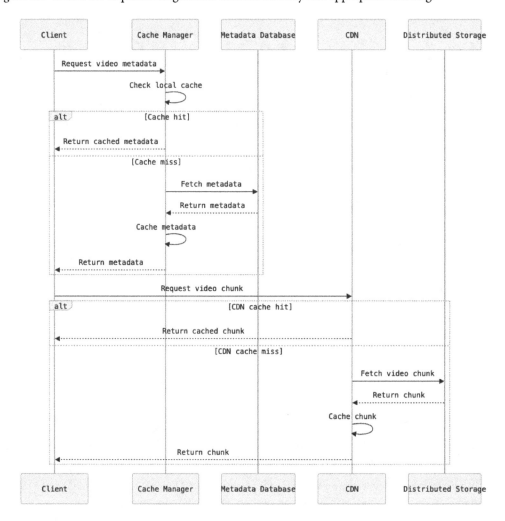

Figure 14.7: The content delivery sequence diagram

Here is the flow:

1. To improve streaming performance and reduce latency, the **Video Service** integrates with a **CDN**.

2. The **CDN** caches video chunks or segments at edge locations closer to the users, reducing the distance between the user and the video content.

3. The **Video Service** can also implement caching mechanisms, such as Redis or Memcached, to store frequently accessed video metadata and reduce the load on the database.

The Video Service plays a vital role in the Netflix-like architecture, enabling efficient video storage, encoding, and streaming. By leveraging distributed storage, encoding workflows, and CDNs, the Video Service ensures a seamless and high-quality video streaming experience for users.

In the next section, we'll explore the User Service, which handles user authentication, profile management, and personalization aspects of our Netflix-like platform.

User Service

The User Service is responsible for managing user authentication, authorization, and profile data in our Netflix-like platform. It handles user registration, login, and profile management functionalities. *Figure 14.8* shows the User service at a very high level.

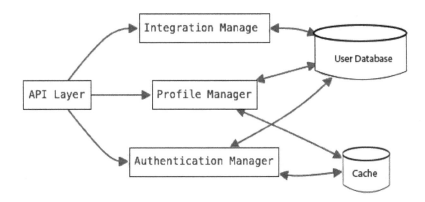

Figure 14.8: The User service

Let's explore the design and implementation details of the User Service. We'll begin by checking the API endpoints of the User service:

- POST /users: Create a new user account:

  - Request body: User information (email, password, name, etc.)

  - Response: User ID and authentication token

- POST /users/login: User login:

  - Request body: User credentials (email and password)

  - Response: An authentication token

- `GET /users/{userId}`: Retrieve user profile information:

  - `Response`: User profile data (name, email, profile picture, etc.)

- `PUT /users/{userId}`: Update user profile information:

  - `Request body`: Updated user profile data

  - `Response`: Updated user profile

- `POST /users/{userId}/profiles`: Create a new profile for a user:

  - `Request body`: Profile information (name, avatar, preferences, etc.)

  - `Response`: A profile ID

- `GET /users/{userId}/profiles`: Retrieve all profiles of a user:

  - `Response`: A list of user profiles

- `PUT /users/{userId}/profiles/{profileId}`: Update a user profile:

  - `Request body`: Updated profile information

  - `Response`: An updated profile

Let's next look at how we can design several workflows for this service, starting with the user authentication and authorization flow.

User authentication and authorization

Figure 14.9 shows the user authentication and authorization flow for Netflix.

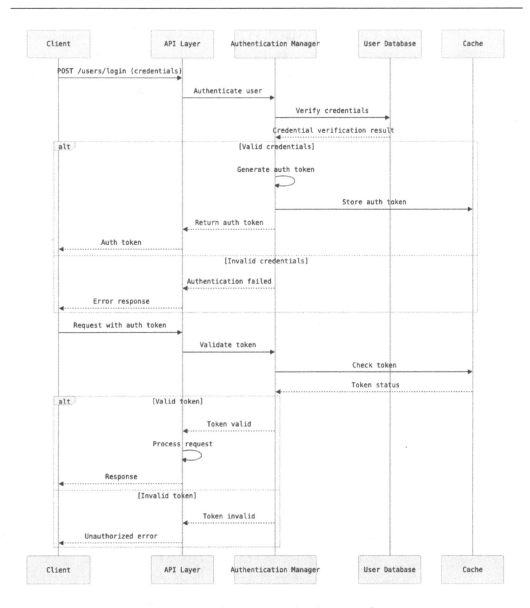

Figure 14.9: Authentication and authorization flow

Here is the flow:

1. When a user registers or logs in, the **User Service** verifies the provided credentials against the user database.

2. Upon successful authentication, the **User Service** generates an authentication token (e.g., JWT) that contains the user ID and other relevant information.

3. The authentication token is returned to the client application, which includes it in subsequent requests to authenticate and authorize the user.

4. The **User Service** validates the authentication token for each incoming request to ensure that the user is authenticated and authorized to access the requested resources.

Next, let's look at the profile management workflow.

User profile management

Figure 14.10 shows the user profile management sequence diagram.

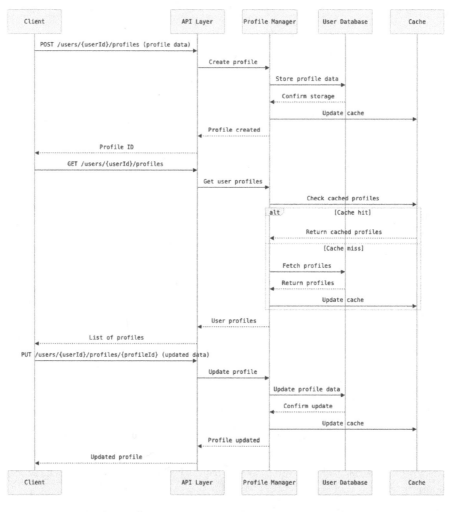

Figure 14.10: The user profile management sequence diagram

Here is the flow:

1. The **User Service** allows users to create and manage multiple profiles within their account.

2. Each profile represents a separate viewing experience, with its own preferences, watch history, and recommendations.

3. The **User Service** provides APIs to **create**, retrieve, update, and delete user profiles.

4. Profile information, such as name, avatar, and preferences, is stored in the user database.

Now that we have looked at some workflows, we should also look at what other services will the user service integrate with.

The User Service integrates with other services in the Netflix-like architecture to provide personalized experiences. It communicates with the Recommendation Service to retrieve personalized video recommendations, based on user preferences and viewing history. The User Service interacts with the Video Service to retrieve user-specific video metadata, such as watch history and favorites. It collaborates with the Billing and Subscription Service to manage user subscriptions and payment information.

Next, we will cover what database, caching, and security are needed for our service.

Database and caching

The User Service uses a relational database, such as PostgreSQL or MySQL, to store user and profile information. The database schema includes tables for users, profiles, preferences, and authentication tokens. To improve performance and reduce database load, the User Service can employ caching mechanisms, such as Redis or Memcached, to store frequently accessed user and profile data.

Let's look at the security and privacy requirements for our service.

Security and privacy

The User Service implements secure authentication and authorization mechanisms to protect user data and prevent unauthorized access. User passwords are hashed and salted before storing them in the database to ensure their confidentiality. Sensitive user information, such as payment details, is encrypted and stored securely. The User Service adheres to data protection regulations and privacy laws, such as GDPR (General Data Protection Regulation) or CCPA (California Consumer Privacy Act), to ensure user privacy and data security.

The User Service plays a crucial role in managing user authentication, authorization, and profile data in our Netflix-like platform. By providing secure and scalable APIs for user and profile management, it enables personalized experiences and seamless integration with other services.

In the next section, we'll explore the Recommendation Service, which generates personalized video recommendations for users, based on their preferences and viewing history.

Recommendation Service

The Recommendation Service is a key component of our Netflix-like platform, responsible for generating personalized video recommendations for users. It analyzes user behavior, preferences, and viewing history to provide relevant and engaging content suggestions. *Figure 14.11* shows at a very high level what services the **Recommendation Service** interacts with.

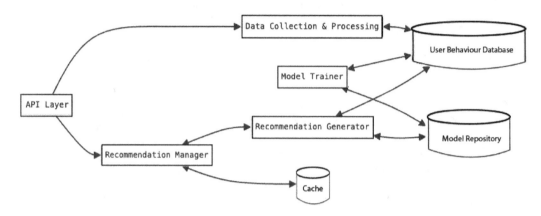

Figure 14.11: The Recommendation service high-level diagram

Let's dive into the design and implementation details of the Recommendation Service. The following are the API endpoints:

- `GET /recommendations/{userId}`: Retrieve personalized video recommendations for a user:

  - `Request parameters`: The user ID, number of recommendations, and filters (e.g., genre or language)

  - `Response`: A list of recommended videos with metadata

- `POST /events`: Record user events for recommendation purposes:

  - `Request body`: The user ID, event type (e.g., the video watched, rated, and searched), and event data

  - `Response`: Success status

Let's now see how the recommendations are generated for a user.

The recommendation generation flow

Figure 14.12 shows the sequence diagram for the recommendation generation flow.

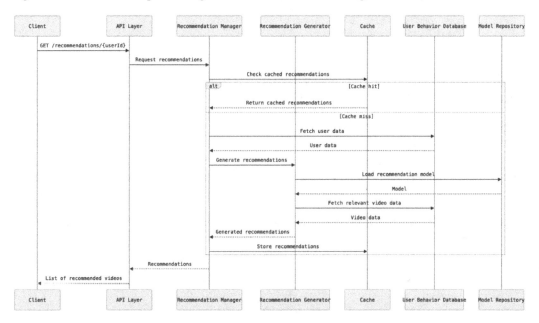

Figure 14.12: The recommendation generation flow

Here is the flow:

1. When a user requests recommendations, the **client** sends a GET request to the /recommendations/{userId} endpoint.

2. The **API layer** forwards the request to the Recommendation Manager.

3. The **Recommendation Manager** first checks the cache for existing recommendations:

    -   If they are found in the cache, it returns the cached recommendations

    -   If they are not found, it proceeds to generate new recommendations

4. For new recommendations, the following occurs:

    I.   The **Recommendation Manager** fetches user data from the User Behavior Database.

    II.   It then requests the **Recommendation Generator** to create personalized recommendations.

    III. The **Recommendation Generator** loads the appropriate model from the **Model Repository**.

    IV.   It then fetches relevant video data from the **User Behavior Database**.

    V.   The generator applies the model to create a list of recommendations.

5.   The **Recommendation Manager** stores the new recommendations in the cache for future quick access.

Finally, the list of recommended videos is returned to the client through the API Layer.

Next, let's see how user events get recorded.

The user event recording flow

Figure 14.13 shows the sequence diagram for the user event recording flow.

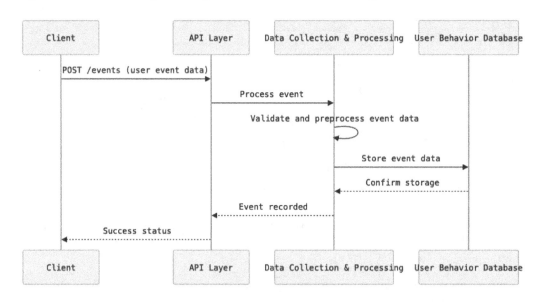

Figure 14.13: The user event recording flow

Here is the flow:

1.   When a user performs an action (e.g., watching a video), the client sends a POST request to the /events endpoint with the event data.

2.   The API layer forwards the event to the data collection and processing component.

3.   The event data is validated and preprocessed.

4.   The processed event data is then stored in the User Behavior Database.

A success status is returned to the client via the API layer.

Next, let's cover the flow for the model training and deployment process.

The model training and deployment process

Figure 14.14 shows the flow for the model training and deployment process.

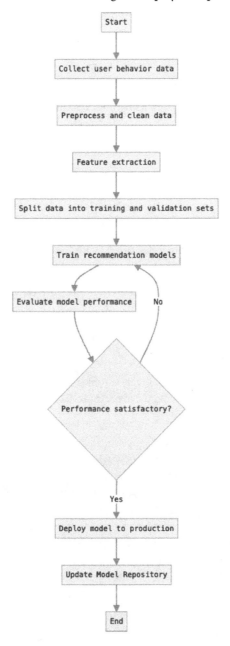

Figure 14.14: The model training and deployment process

Here is the flow:

1. The process begins with collecting user behavior data from the **User Behavior Database**.

2. The data is preprocessed and cleaned to remove any inconsistencies or errors.

3. Relevant features are extracted from the cleaned data.

4. The dataset is split into training and validation sets.

5. Various recommendation models are trained using the training data.

6. The models' performance is evaluated using the validation set.

7. If the performance is satisfactory, the best-performing model is deployed to production.

8. The **Model Repository** is updated with the new model.

9. If the performance is not satisfactory, the process returns to the model training step with adjusted parameters.

Now, let's look at the Recommendation Service's integration with other services:

- The Recommendation Service integrates closely with the User Service to retrieve user profiles and viewing history

- It communicates with the Video Service to fetch up-to-date video metadata, ensuring accurate recommendations

The service also interacts with the Analytics and Reporting Service, providing performance metrics and insights about the recommendation system

These flows and processes work together to create a robust and efficient Recommendation Service. Let's now cover the scalability and performance aspects of this service.

Scalability and performance

The Recommendation Service is designed to handle a large number of concurrent recommendation requests. It can be scaled horizontally by deploying multiple instances of the service behind a load balancer, distributing the incoming traffic. Caching mechanisms, such as Redis or Memcached, can be employed to store frequently accessed recommendation results and improve response times. The recommendation models can be deployed on scalable infrastructure, such as Apache Spark or TensorFlow Serving, to handle high-volume requests efficiently.

Next, let's look at monitoring and evaluation.

Monitoring and evaluation

The Recommendation Service integrates with monitoring systems to track key metrics and performance indicators. Metrics such as recommendation request latency, cache hit rate, and model accuracy are monitored to ensure the health and effectiveness of the service. A/B testing and online evaluation techniques can be employed to assess the impact of different recommendation algorithms and configurations on user engagement and satisfaction. User feedback and explicit ratings can be collected to evaluate and improve the quality of recommendations over time.

The Recommendation Service is a critical component of our Netflix-like platform, enabling personalized and engaging video recommendations for users. By leveraging advanced recommendation algorithms, data collection, and real-time processing, it enhances user satisfaction and retention.

In the next section, we'll explore the CDN, which plays a vital role in efficiently delivering video content to users across the globe.

The CDN

The CDN is a crucial component of our Netflix-like platform, responsible for efficiently delivering video content to users across the globe. It ensures high performance, scalability, and reliability by distributing video content across a network of geographically dispersed servers. *Figure 14.15* covers the CDN architecture.

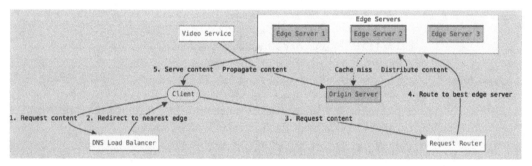

Figure 14.15: The CDN architecture

Let's explore the design and implementation details of the CDN.

CDN architecture and content distribution

Figure 14.16 covers the flow for the content distribution sequence diagram

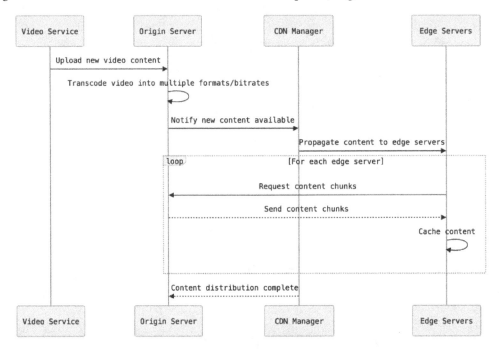

Figure 14.16: The content distribution sequence diagram

Here is the flow:

1. The CDN consists of a network of edge servers located in various regions around the world, connected to an origin server that holds the master copy of all content.

2. When new video content is uploaded to the platform via the Video Service, it triggers the following process:

    *   The origin server transcodes the video into multiple formats and bitrates to support different devices and network conditions

    *   The CDN manager is notified of the new content availability

    *   The CDN manager initiates content propagation to the edge servers

    *   Each edge server requests and caches the content chunks from the origin server

3. This distribution ensures that content is readily available, closer to the users, reducing latency and improving streaming performance.

Next, let's look at the request routing and video streaming workflow.

Request routing and video streaming

Figure 14.17 shows the sequence diagram for request routing and video streaming.

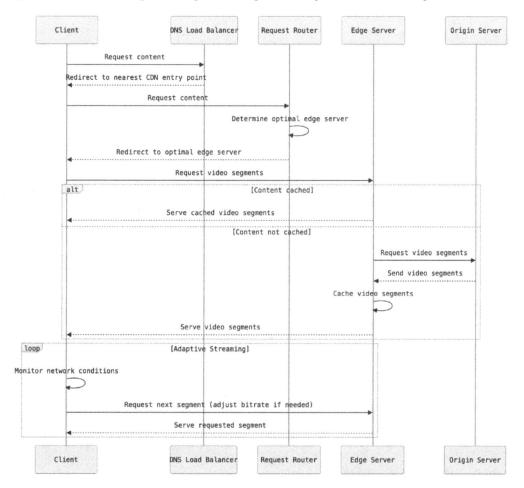

Figure 14.17: The request routing and video streaming workflow

Here is the flow:

1. When a user requests to stream a video, the client application initiates the following sequence:
2. The **client** sends a request to the **DNS Load Balancer**.
3. The **DNS Load Balancer** redirects the client to the nearest CDN entry point.

4. The client then contacts the **Request Router**.

5. The **Request Router** analyzes factors such as user location, server availability, network latency, and server load to determine the optimal edge server.

6. The **client** is redirected to the selected edge server.

- The edge server then serves the video content to the user:

    - If the content is cached, it's served directly from the edge server

    - If not cached, the edge server retrieves the content from the origin server, caches it, and then serves it to the client

This intelligent routing ensures a fast and reliable streaming experience for users worldwide.

In order to support compelling user experiences, our Netflix-like system must support adaptive bitrate streaming, which we will discuss next.

Adaptive bitrate streaming

The CDN supports adaptive bitrate streaming to provide optimal video quality:

- Video content is encoded into multiple bitrates and segmented into small chunks.

- During playback, the client continuously monitors network conditions.

- The client requests subsequent video segments from the edge server, adjusting the bitrate as needed, based on network performance.

- The edge server serves the requested segments, allowing for seamless quality adjustments during playback.

- This adaptive approach ensures a smooth streaming experience across varying network conditions and device capabilities.

Let's now talk about how our CDN design deals with content security.

Content security and DRM

The CDN integrates with a (**Digital Rights Management**) **DRM** Service to implement content security measures:

- Video content is encrypted using DRM systems, such as Microsoft PlayReady or Google Widevine

- The CDN collaborates with the DRM Service for secure content delivery and license management

- Only authorized users with valid DRM licenses can access and play the protected video content

This integration ensures content protection against unauthorized access and piracy while complying with licensing agreements and regional restrictions

The CDN seamlessly integrates with other components of our Netflix-like platform:

- **Video Service**: Facilitates content propagation and retrieval, ensuring efficient video transcoding, storage, and distribution

- **Recommendation Service**: Enables personalized content delivery, allowing for dynamic and personalized playlists or recommendations

- **Analytics Service**: Provides comprehensive insights and metrics on content delivery performance, helping to optimize the CDN's operations

- **DRM Service**: Ensures secure content delivery through encryption and license management

This integrated approach allows the CDN to not only deliver content efficiently but also contribute to a personalized and secure user experience. By leveraging this sophisticated CDN architecture and its integrations, our Netflix-like platform can deliver high-quality video content to users worldwide with minimal latency, adaptive quality, and robust security, ensuring an excellent streaming experience for our global audience.

Summary

In this chapter, we explored the system design of a Netflix-like platform, covering various aspects such as functional and non-functional requirements, data modeling, capacity estimation, high-level architecture, and the design of key components, such as the Video Service, User Service, Recommendation Service, and CDN.

We started by defining the core features and functionalities that our platform should support, ensuring a comprehensive and user-centric streaming experience. We then discussed the non-functional requirements, emphasizing the importance of scalability, performance, reliability, and security.

To handle the massive amount of data and traffic generated by millions of users, we delved into data modeling and capacity estimation. We designed a data model that captures the essential entities and relationships, and we performed calculations to estimate the storage, bandwidth, and processing requirements of our platform.

Based on these requirements and estimations, we proposed a high-level architecture that incorporates various components and services, ensuring a modular and scalable design. We explored the responsibilities and interactions of each component, focusing on efficient video storage, encoding, streaming, recommendation generation, and content delivery.

Throughout the chapter, we emphasized the importance of scalability, fault tolerance, and performance optimization. We discussed techniques such as data sharding, caching, content distribution, and adaptive bitrate streaming to ensure a seamless and reliable streaming experience for users. We have discussed many different services that present different learnings of designing scalable systems. In the next chapter, we will focus on some tips to interview for System Design based on our experience interviewing candidates and talking to Senior Engineers at several Fortune 100 companies..

15

Tips for Interviewees

System design interviews are a crucial part of the hiring process for senior software engineering and engineering management roles. They test your ability to architect scalable, efficient, and robust systems, reflecting your understanding of the trade-offs and complexities involved in real-world applications.

Performing well in these interviews isn't only required to get an offer. Candidates are often evaluated on their performance on these system design rounds to be considered for different levels. For example, a candidate who performs excellently at these system design interviews may be hired at an L+1 level rather than an L level. That means higher responsibilities, scope, and compensation.

This chapter offers practical tips and strategies to help you excel in system design interviews, from understanding the requirements to presenting your solution effectively.

We will cover the following topics in this chapter:

- Tips for preparation for system design interviews
- Tips for the interview session

Tips for preparation for system design interviews

Preparing for system design interviews requires a strategic approach that combines theoretical knowledge, practical experience, and effective communication skills. This chapter provides a comprehensive guide on how to prepare for system design interviews, covering essential topics, resources, and techniques to help you succeed.

Understanding the fundamentals

Start with a solid understanding of the fundamental concepts in system design. These foundational basics will help you navigate the path and solve the core challenges, especially when faced with an unfamiliar system design question. With a solid grasp of these concepts, you will be able to perform better in your interviews:

- **Scalability**: Learn about horizontal and vertical scaling, load balancing, and distributed systems. Learn how load balancers allocate incoming network traffic among multiple servers to prevent any single server from becoming a bottleneck.

- **Databases**: Familiarize yourself with different types of databases (SQL versus NoSQL), indexing, sharding, and replication. Know when to use a relational database such as PostgreSQL versus a NoSQL database such as MongoDB based on the use case.

- **Caching**: Understand caching strategies, cache eviction policies, and tools such as Redis and Memcached. Use Redis to cache frequently accessed data to reduce database load and improve response times.

- **Consistency and availability**: Study the CAP and PACELC theorems and learn about consistency models (strong, eventual consistency). Understand the trade-offs between consistency and availability in distributed systems.

This is not a comprehensive list, but if you have read the first couple of chapters of this book, you would already have gone through these and more. Let's move on to learning about some of the common system design patterns that will come in handy in a system design interview.

Studying common system design patterns

Learning about common system design patterns and their applications will help you to have a good starting point for your solution. Some of these are as follows:

- **Microservices architecture**: Understand how to design systems using microservices, including inter-service communication. Use RESTful APIs or gRPC for communication between microservices.

- **Event-driven architecture**: Learn how to design systems that respond to events using message queues or event streams. Use Apache Kafka to build a real-time data processing pipeline.

- **Service-Oriented Architecture (SOA)**: Study how SOA differs from microservices and its use cases. Use SOAP for communication in legacy systems or when strong contract enforcement is required.

- **Design patterns**: Familiarize yourself with patterns such as **Command Query Responsibility Segregation (CQRS)**, Saga, and Circuit Breaker. Implement the Circuit Breaker pattern to handle failures in distributed systems gracefully.

Again, this is not a comprehensive list, but identifying and mastering such design patterns is very important for your system design interview preparation.

Practicing designing systems

Practical experience is essential to mastering system design. Regularly practicing the design of systems for various scenarios helps solidify theoretical knowledge and improves problem-solving skills. Engaging with diverse use cases, from designing scalable web applications to architecting robust databases, equips you with the versatility needed to tackle real-world challenges. Here are some steps to consider:

- **Mock interviews**: Conduct mock interviews with peers or use platforms such as Pramp, Interviewing.io, or Exponent. Simulate a system design interview scenario, such as designing a URL shortening service, and get feedback on your approach.

- **Design challenges**: Take on design challenges from resources such as LeetCode, HackerRank, or *Grokking the System Design Interview*. Solve design problems such as building a scalable notification system or a ride-sharing service.

- **Case studies**: Analyze real-world case studies of large-scale systems to understand their architecture and design decisions. Study the architecture of Twitter to learn about handling high throughput and ensuring low latency.

Consistent practice and review not only enhance your understanding of design principles but also build confidence in your ability to create efficient, reliable, and maintainable systems.

Learning from online resources

You can utilize online resources and courses to deepen your understanding and stay updated with best practices. Here is a list of some of these resources:

- **Books**: Read foundational books such as *Designing Data-Intensive Applications* by Martin Kleppmann and *The Art of Scalability* by Martin L. Abbott and Michael T. Fisher. These books provide in-depth knowledge on building scalable and reliable systems.

- **Online courses**: Enroll in online courses from platforms such as Coursera, Udemy, or Educative. *Grokking the System Design Interview* on Educative offers practical insights and design patterns.

- **Blogs and articles**: Follow blogs and articles from industry leaders and companies known for their robust architectures. Read engineering blogs from companies such as Netflix, Uber, and LinkedIn to learn about their system design practices.

- **YouTube channels**: Watch videos and tutorials on system design topics. There are many great channels, such as Tushar Roy, Tech Dummies, and Gaurav Sen, which provide detailed explanations of system design concepts and interview tips.

System design is a vast and complex field. It's important to keep learning and look at problems and solutions from different perspectives.

Honing your communication skills

Effective communication is key to conveying your ideas clearly during the interview:

- **Practice verbalizing your thoughts**: Regularly practice explaining your thought process out loud. Conduct mock interviews with a friend or mentor and focus on articulating your design decisions clearly.

- **Use visual aids**: Use diagrams and charts to illustrate your design. Practice drawing architecture diagrams and sequence diagrams to convey complex ideas visually.

Reviewing and reflecting

After each practice session or real interview, review your performance and identify areas for improvement:

- **Self-assessment**: Reflect on what went well and what could be improved. Did you manage to cover all the requirements? Were there any components you overlooked?

- **Seek feedback**: Get feedback from peers, mentors, or interviewers to understand your strengths and weaknesses. Ask for specific feedback on your approach, communication skills, and depth of knowledge.

- **Continuous improvement**: Use the feedback to refine your preparation strategy and focus on areas that need improvement. For example, if you struggled with database design, spend more time studying database architectures and practicing related problems.

Preparing for system design interviews requires a combination of theoretical knowledge, practical experience, and effective communication skills. By understanding the fundamentals, studying common design patterns, practicing regularly, adopting a methodical approach, leveraging online resources, honing your communication skills, and reflecting on your performance, you can build the confidence and expertise needed to excel in system design interviews. Continuous learning and improvement are key to staying ahead and successfully navigating the complexities of system design challenges.

Tips for the interview session

Now that you have done extensive preparation, when it comes to the actual interview, you want to keep some tips in mind. First of all, be prepared that the questions you are asked may be unfamiliar to you. That's OK. Don't panic. Most of the time you should be able to use the fundamental concepts, components, philosophies, techniques, and subsystems you have learned about and designed to design a new system. Let's go over some guidelines and tips now.

Understanding the problem statement

Before diving into designing the system, take the time to thoroughly understand the problem statement:

- **Ask clarifying questions**: Ensure you have a clear understanding of the requirements. Ask questions about the scope, constraints, and any ambiguities. For example, if asked to design a chat application, clarify whether the focus is on real-time messaging, user authentication, or message storage.

- **Identify functional requirements**: List the core functionalities the system must support. For example, for an e-commerce platform, these might include user registration, product search, shopping cart, and payment processing.

- **Identify non-functional requirements**: Understand performance, scalability, availability, and reliability requirements. For example, for a social media feed, you might need to handle high read and write throughput with low latency.

Breaking down the problem

Divide the problem into smaller, manageable components or services:

- **Component identification**: Identify the major components or services required. For example, in a URL shortening service, the components might include an API gateway, URL storage, redirection service, and analytics.

- **Define interactions**: Determine how these components will interact with each other. For example, the API gateway receives shortening requests, the storage service saves the mapping, and the redirection service handles redirects.

- **Use diagrams**: Visual aids such as block diagrams or sequence diagrams can help convey your design clearly. For example, draw an architecture diagram showing how data flows between the user, API gateway, storage, and redirection service.

Key steps to follow

A very high-level set of steps you should take are as follows:

1. Write functional requirements:

   - Write key functional requirements and refrain from spending a lot of time brainstorming and thinking about new features and auxiliary use cases

   - Be focused and obtain clarification from the interviewer on the requirements

2. Jot down the non-functional requirements:

   - One simple way is to just enumerate all the common non-functional requirements and see whether it's relevant to the problem and scope.

   - There are a couple of important points to remember here. Don't just say general phrases. For example, don't just say the system should be highly available, but talk about specifics, such as 99.9% or 99.99% availability. Similarly, saying that the system should be highly consistent is a very general statement. Instead, talk about consistency levels for specific sub-use cases.

3. List out the APIs:

   - Note down the customer-facing and internal system APIs needed for the functional requirements to be satisfied

   - Preferably you should use REST APIs, but you could use just functions/ methods to keep it simple

4. Do the high-level calculations and estimates:

   - Make sure the estimates are purposeful – meaning that the calculations should influence your design choices.

   - Use round numbers closer to powers of 10s, so calculations are easier. For example, 86,400 seconds in a day can be approximated to 100,000 seconds in a day for easier calculations.

5. Create a high-level block diagram:

   - Draw a high-level initial block diagram with the major components, such as client device, load balancer, app server, microservices, and databases.

   - This will ensure that you have a starting point and the architecture diagram will inform you of the single point of failure or the major choking points. This is how you would identify what the bottlenecks and core challenges are.

6. Address the core challenges:

   - Address the bottlenecks and challenges to refine your design

   - Make sure you are listening to clues provided by your interviewer and that they are on board to do deep dives into the areas you have identified

7. Draw the final high-level architecture diagram and flow after making the changes to the initial high-level block diagram.

8. Wrap it up by verifying that all the functional and non-functional requirements are satisfied with your final design.

Communicating your solution effectively

As an interviewer for over a decade, I've observed that presenting your solution clearly is as important as the solution itself. I recall many candidates who had brilliant designs for a scalable system. However, their inability to articulate their ideas effectively made it difficult for the panel to fully grasp the strengths of their approach. Conversely, I have seen many other candidates with much simpler designs excelling because they communicated their thought processes and design decisions with clarity, using diagrams and concise explanations.

So here are some tips for communicating your solution better:

- **Structure your presentation**: Follow a logical structure – state the problem, outline your high-level approach, dive into components, and discuss trade-offs. Start with the system's objectives, then describe the architecture, followed by component details and how they interact.

- **Highlight key decisions and explain trade-offs**: Explain the reasoning behind your choices. Justify why you chose a particular database or caching strategy based on the requirements. This can be a key differentiator between a senior and a junior-level candidate.

- **Listen to the hints and clues provided by the interviewer**: It's a collaborative exercise, and your interviewer is collecting all the signals needed for them to write up your evaluation. If they are steering the conversation to a particular set of problems, please listen to them.

- **Adapt and iterate based on feedback**: Be flexible and willing to adapt your design based on the interviewer's feedback. Treat the design process as iterative, refining your approach as new information or feedback is received.

- **Communicate verbally as well as in writing**: Write down your thoughts and take advantage of diagrams, blocks, and flows to explain your thoughts and solution direction. Learn how to speak and write or draw at the same time. Use all the tools and channels to enhance communication.

- **Summarize your design**: Conclude with a summary of your design, reiterating key points and decisions. Briefly recap the high-level architecture and main components. Emphasize the strengths of your design and acknowledge any trade-offs or limitations.

- **Prepare for follow-up questions**: Be ready to answer follow-up questions or delve deeper into specific areas. Be prepared to discuss details about components, algorithms, and design decisions. Expect questions on how your design handles scaling and failure scenarios.

- **Leverage your experience**: Use your past experience to inform your design decisions and demonstrate your expertise. Draw parallels between the problem at hand and similar projects you've worked on. Share insights and lessons learned from past projects, including what worked well and what didn't.

- **Keep the other requirements in mind**: Consider requirements such as the reliability, observability, debuggability, and usability of the system design and share your views at the appropriate time, usually at the end, to demonstrate that you care about these as well.

The ability to present one's solution is crucial, as it not only demonstrates technical understanding but also showcases one's capability to collaborate and convey complex concepts to others. This can be a deal-breaker in an interview and many times decides the seniority or the leveling at a company.

Summary

In this chapter, we learned that system design interviews are crucial for senior software engineering and engineering management roles, assessing your ability to architect scalable, efficient, and robust systems. Excelling in these interviews not only helps you secure job offers but also potentially positions you for higher-level roles with greater responsibilities and compensation.

We looked at some tips for preparing for the interviews. To prepare effectively, start by understanding the fundamentals of system design. This includes concepts such as scalability, databases, caching, and consistency models. Familiarize yourself with common design patterns, such as microservices, event-driven architectures, and SOA. Practicing system design through mock interviews and design challenges and analyzing case studies of real-world systems is also essential.

We then moved on to leveraging online resources, including foundational books, courses, blogs, and YouTube channels, to deepen our understanding and stay updated with best practices. We learned how honing your communication skills by regularly practicing how to articulate your design decisions clearly and using visual aids such as diagrams and charts can help you prepare better.

Then, we moved on to the tips that you should keep in mind while the interview is going on. During the interview, focus on thoroughly understanding the problem statement by asking clarifying questions and identifying both functional and non-functional requirements. Break down the problem into manageable components, define interactions, and use visual aids to convey your design. Communicate your solution effectively, structure your presentation logically, highlight key decisions, and be prepared to adapt based on feedback from the interviewer.

Lastly, we concluded this chapter by sharing how by combining theoretical knowledge, practical experience, and effective communication, you can excel in system design interviews. Continuous learning and improvement are key to staying ahead and successfully navigating the complexities of system design challenges.

In the next chapter, we will share a system design cheat sheet with a lot of quick patterns and insights that you can brush up on to ace your technical interviews.

16
System Design Cheat Sheet

Welcome to this final chapter, a cheat sheet designed to equip you with the essential strategies to ace your technical interviews. This chapter is meticulously crafted to provide structured insights into key aspects crucial for mastering system design assessments. Whether you're gearing up for your next interview or seeking to enhance your system architecture skills, this chapter offers practical solutions to common questions that arise during system design interviews.

Throughout this chapter, we will explore the structured approach necessary to excel in system design interviews. From understanding how to effectively clarify problem statements and outline functional and non-functional requirements to creating high-level architectural diagrams, each step is designed to ensure you approach interviews with confidence and clarity.

Additionally, we will delve into critical decisions such as selecting the optimal data store based on use case requirements, choosing the right data structures to maximize efficiency, and identifying the most suitable components and protocols for various system challenges. By the end of this chapter, you'll be equipped with a comprehensive toolkit to navigate and excel in any system design interview scenario.

The chapter covers the following core questions:

- What structure should we follow in a system design interview?
- Which data store should we use for a use case?
- Which data structures should we use for a use case?
- Which components should we use for which use case?
- What protocol should we use for which use case?
- Which solution should we use for which core challenge?

Let's jump right in.

What structure should we follow in a system design interview?

A system design interview is a comprehensive evaluation of your ability to architect scalable, reliable, and maintainable systems. Here's a structured approach to follow during a system design interview to ensure you cover all essential aspects:

1. Ask and clarify the problem

2. List out the functional requirements

3. List out the non-functional requirements

4. Write down the APIs

5. Do high-level estimates and calculations

6. Draw a high-level system design diagram addressing the functional requirements without focusing too much on an optimized solution

7. Identify core challenges and address them by brainstorming various options and making the right trade-offs

8. Put together a final high-level system design and architecture

9. Verify the functional and non-functional requirements

Now that we know the high-level steps to be followed in an interview, let's explore some core questions and their answers.

Which data store should we use for a use case?

Choosing the right data store for a specific use case depends on various factors, including the nature of the data, access patterns, scalability requirements, consistency, latency, and the overall architecture of the application. Here are some guidelines and examples to help you select the most appropriate data store for different scenarios:

Use case	Data store
<ul><li>Structured data</li><li>Require ACID properties</li><li>Not sparse and not a huge number of rows</li><li>Not a lot of joins else reads will be slow</li></ul>	<ul><li>Relational database (MySQL, PostgreSQL, or Oracle)</li><li>Shard to support scale</li></ul>
<ul><li>Non-structured, very high scale</li><li>Wide variety of documents, such as Amazon items (sparse data)</li><li>Data is finite</li></ul>	<ul><li>Document database (MongoDB or Couchbase)</li></ul>

Use case	Data store
• Non-structured, very high scale • Ever-increasing data • Large volume of data with thousands of columns • Most of the time just queries only a few columns	• Columnar databases • Hbase (consistency over availability) • Cassandra (availability over consistency, tunable consistency)
Scalable and fast key-value store	Redis or Memcached
Fast free-text search	Lucene, Elasticsearch, or Solr
Fast writes	WAL
Fast reads	Caching, replications, in memory, CDNs
Blob store video and images	S3/CDNs
Complex relations such as in a graph	Graph db (Neo4j)
Hot data	In memory, SSDs
Cold data	Disk, Amazon Glacier
Find "highly similar" data in a set of unstructured data (such as images, text blobs, and videos). This is particularly needed in AI applications.	Vector database
Time-series metrics data	Time series (OpenTSDB)
Proximity or nearby entity search	Geo-spatial index (quadtrees or geohashing)

Table 17.1: Choosing the right data store for different use cases

The preceding table lists the use cases for data store mapping. Now, let's go over the data structures to be used for different use cases in the next section.

Which data structures should we use for a use case?

Choosing the right data structure for a use case depends on the specific requirements and constraints of the problem you are trying to solve. Here is a guide on choosing the right data structure for different use cases:

Example use cases	Data structure
• Find whether an element is a member of a set when the space efficiency and speed of query operations are critical, even at the expense of a small probability of false positives. • Ensure that a web cache does not store duplicate URLs. • Check whether a key might be in a database table before performing a costly disk access. • Filter out known spam emails efficiently and check whether an incoming email matches any email in a database of known spam addresses	Bloom filter
• Estimate the frequency of elements in a data stream in a space-efficient way, handling large-scale data streams with fixed memory usage. • Monitor and analyze network traffic to detect heavy hitters or frequent items by keeping track of the frequency of packets or flows. • Monitor large-scale social media activity or website traffic by tracking the frequency of events in real time, such as clicks, views, or transactions. • Suggest popular or trending items to users by maintaining approximate counts of item views, purchases, or ratings.	Count min sketch
• Approximate the cardinality (i.e., the number of distinct elements) of a multiset with high accuracy and low memory usage. • Measure the reach and effectiveness of advertising campaigns without storing detailed logs by estimating the number of unique users who have seen or clicked on an ad. • Understand network usage patterns and detect anomalies such as DDoS attacks by counting unique IP addresses, sessions, or flows in network traffic data. • Measure the reach and impact of social media campaigns by estimating the number of unique users liking, sharing, or commenting on posts.	Hyper log log

Example use cases	Data structure
• Efficiently verify data integrity with fingerprinting data, allowing you to confirm whether the data has been tampered with without needing to download the entire dataset. • Git version control system: Track changes to code efficiently. Each commit in Git history has a unique Merkle root representing the state of the code base at that point. This allows developers to verify the integrity of specific versions and identify changes made over time. • P2P file sharing: Ensure downloaded files are complete and unaltered. The file is divided into chunks, and each chunk is hashed. The complete file's Merkle root is distributed along with the chunks. Anyone downloading the file can verify its integrity by checking the hashes of the downloaded chunks against the Merkle root. • Software updates: Merkle trees can be used to ensure the downloaded file is complete and hasn't been corrupted during transmission. The update provider can publish the Merkle root of the file beforehand, and users can verify the downloaded file's integrity by calculating its Merkle root and comparing it to the published one.	Merkle tree

Table 17.2: Choosing the right data structure for different use cases

In the next section, let's go through the components to be selected for different use cases.

Which components should we use for which use case?

In system design, choosing the right components is crucial to building a scalable, reliable, and maintainable system. The following is a guide to help you decide which components to use for various use cases:

A traffic director for the network or application distributing incoming traffic evenly across multiple servers in a pool. • **Web applications**: Distributing traffic across multiple web servers to handle high user volumes for e-commerce sites, social media platforms, or any web application with fluctuating traffic. • **Database clusters**: Balancing read/write requests across multiple database servers in a cluster for improved performance and redundancy.	**Load balancer**

An intermediary between your device and the internet to protect your device. • **Privacy and anonymity**: Proxies can hide your IP address, making it seem like you're browsing from a different location. This is useful for accessing content restricted by geography (content blocked in your region) or for maintaining a level of anonymity online. • **Security**: Some proxy servers offer additional security features, such as filtering out malicious content or encrypting your traffic. This can be helpful when using public Wi-Fi networks where your connection might be less secure. • **Content filtering**: Organizations or schools might use proxies to restrict access to certain websites or types of content (e.g., gambling sites or social media).	Proxy
A middleman between the internet and your web application to protect servers. • **E-commerce websites**: Reverse proxies can handle high volumes of traffic during sales or peak seasons, distributing load and ensuring a smooth shopping experience. • **Content Delivery Networks (CDNs)**: CDNs often use reverse proxies to cache content at geographically dispersed edge locations, bringing content closer to users for faster loading times. • **Microservices architectures**: Reverse proxies can route requests to the appropriate microservice based on specific criteria, simplifying traffic management in complex systems.	Reverse proxy
Manage traffic and protect systems in software design to prevent overload and ensure fair access for everyone. • **API protection**: Limiting the number of API calls an application can make per minute prevents abuse and protects against **Denial-of-Service (DoS)** attacks. • **Login attempts**: Limiting login attempts deters brute-force attacks where someone tries to guess a password repeatedly. • **E-commerce transactions**: Rate-limiting purchase attempts can prevent fraudulent activity or overwhelm payment processing systems during sales.	Rate limiter

Automatic switch that monitors the health of a service or resource, protecting the system from cascading failures. • **Microservices architecture**: In a system with multiple interconnected services, a circuit breaker can isolate a failing service and prevent it from bringing down the entire system. • **External APIs**: If an external API you rely on is experiencing problems, a circuit breaker can prevent your system from constantly retrying failed requests. • **Third-party integrations**: When integrating with a third-party service, a circuit breaker can prevent your system from crashing due to temporary outages with the external service.	Circuit breaker
A central hub for managing all incoming API requests in a microservices architecture. • **Single entry point**: The gateway acts as a single point of entry for all API requests, simplifying client interactions and reducing the need for clients to know the specifics of each backend service. • **Request routing**: The gateway receives requests and routes them to the appropriate backend service based on predefined rules (such as path, headers, or parameters). • **Security**: The gateway can enforce authentication and authorization policies, ensuring only authorized users can access specific functionalities. It can also handle tasks such as encryption and rate limiting to protect backend services. • **Monitoring and analytics**: The gateway can monitor API traffic, track usage patterns, and provide valuable insights into how APIs are being used.	API gateway

A communication channel between systems, making them decoupled so that they don't have to wait for each other to complete tasks: • **E-commerce order processing**: When a customer places an order, a message can be sent to a queue. A separate worker service can then consume the message, process the order (payment, inventory check, and shipping), and update relevant systems asynchronously. This avoids blocking the user interface while order processing happens in the background. • **Task queues**: Long-running tasks, such as video encoding or image processing, can be added to a message queue. Worker services can then pick up these tasks and complete them asynchronously, freeing up the main application to handle other user requests. • **Social media feeds**: When a user follows someone on a social media platform, a message queue can be used to notify them about new posts. The queue stores updates and a separate service can deliver them to the user's feed asynchronously, improving performance and scalability to handle large user bases.	Message queues
Geographically distributed network of servers that work together to deliver content to users with faster loading times and improved user experience. • **Websites and web applications**: Most popular websites and web applications leverage CDNs to ensure fast loading times for users worldwide. This is especially crucial for e-commerce sites where slow loading times can lead to lost sales. • **Streaming services**: Video and music streaming services rely heavily on CDNs to deliver high-quality content with minimal buffering, even during peak usage periods. • **Social media platforms**: Social media platforms with massive user bases utilize CDNs to deliver images, videos, and other content efficiently, ensuring a smooth user experience.	CDNs

Table 17.3: Choosing the right component for different use cases

In the next section, we will learn about various protocols and compare them with each other. Also, we will list several use cases and the applicable protocols to be used for them.

What protocol should we use for which use case?

Before we jump into the different use cases and which protocol to use, let's understand the differences between the different protocols and compare them along the following dimensions:

Feature	HTTP	SSE	WebSockets
Communication model	Request-response	Unidirectional (server to client)	Bidirectional
Connection type	Short-lived	Long-lived	Long-lived
Data format	Text (HTML, JSON, etc.)	Text (event-stream)	Text and binary
Use case examples	Web pages, REST APIs	Live updates, notifications	Real-time chat, games, financial tickers
Latency	High	Low	Very low
Scalability	High (for stateless requests)	Moderate	Moderate (requires careful management)
Automatic reconnection	No	Yes	Application managed
Protocol overhead	High (repeated handshakes)	Low	Low
Browser support	Universal	Modern browsers	Modern browsers

Table 17.4: Different protocols and their comparison table

In the following table, let's explore different use cases and the right protocol to be used along with the rationale for the choice:

Use Case	Recommended Protocol	Rationale
Static content delivery	HTTP	Simple request-response model, well supported
RESTful APIs	HTTP	Stateless, widely used for backend communication
Form submissions	HTTP	Standard way to handle form data
File transfers	HTTP	Efficient for large file uploads/downloads
Real-time notifications	SSE	Simple, automatic reconnection, server-to-client updates

Use Case	Recommended Protocol	Rationale
Live feeds	SSE	Efficient for streaming live updates
Monitoring dashboards	SSE	Continuous updates from server to client
Chat and messaging (simple)	SSE	Simple unidirectional message updates
Online gaming	WebSockets	Low-latency, bidirectional communication
Real-time chat	WebSockets	Fast, continuous message exchange
Collaborative tools	WebSockets	Real-time, bidirectional data exchange
Financial applications	WebSockets	Real-time updates, low latency
IoT applications	WebSockets	Continuous data exchange between devices and servers

Table 17.5: Choosing the right protocol for different use cases

In the next section, let's explore various solutions for different use cases.

Which solution should we use for which core challenge?

Identifying the right solution for specific core challenges is essential to building robust, scalable systems. Each challenge requires tailored strategies and technologies. This section delves into the best solutions for various core challenges, helping you make informed decisions to optimize your system's performance and reliability:

Core Challenge	Description	Potential Solutions
Handling high write throughput	Managing systems with very high write rates (e.g., logging, real-time analytics)	• Write-Ahead Logging (WAL) • Sharding • NoSQL databases
Ensuring data consistency	Maintaining consistency in a distributed system	• Distributed transactions (e.g., two-phase commit) • Eventual consistency • Conflict resolution

Core Challenge	Description	Potential Solutions
Low-latency requirements	Providing responses with minimal delay	• In-memory databases (e.g., Redis) • Caching (e.g., Memcached) • Edge computing
Scalability	Scaling the system to handle increased load	• Horizontal scaling • Load balancing • Microservices architecture
Fault tolerance	Ensuring the system continues to operate despite failures	• Replication • Failover mechanisms • Circuit breakers
Data partitioning	Distributing data across multiple nodes	• Hash partitioning • Range partitioning • Consistent hashing
Search performance	Providing fast and relevant search results	• Inverted index • Search engines (e.g., Elasticsearch) • Caching
Handling spiky traffic	Managing sudden spikes in traffic (e.g., sales or events)	• Autoscaling • Load smoothing (e.g., request queuing or rate limiting) • CDNs
Distributed locking	Coordinating access to shared resources in a distributed system	• Distributed Lock Managers (DLMs) • ZooKeeper • Redis

Table 16.6: Choosing the right solution idea for different use cases

Summary

In this chapter, we explored a structured approach to excel in system design interviews, focusing on problem clarification, requirement listing, API design, and architectural diagramming. We've also delved into critical decisions such as selecting appropriate data stores, choosing optimal data structures, and identifying suitable components and protocols to tackle core system challenges. By providing practical insights and guidelines, this chapter equips you with the necessary tools to confidently navigate system design interviews and build scalable, reliable, and maintainable systems.

This concludes the last chapter of this book. We hope that this system design book serves as a comprehensive guide to mastering the art of architecting scalable, reliable, and maintainable systems. Covering essential topics and deep dives, it equips you with the knowledge and skills needed not only for successful system design but also to excel at interviews.

As you move forward, remember that practical experience is crucial. Regularly practice designing systems for different scenarios, analyze real-world case studies, and engage in mock interviews. Stay updated with the latest trends and best practices by leveraging online resources and honing your communication skills.

Embrace the journey of continuous learning and improvement, and approach each challenge with confidence and curiosity. We wish you all the very best!

Index

S

packtpub.com

Subscribe to our online digital library for full access to over 7,000 books and videos, as well as industry leading tools to help you plan your personal development and advance your career. For more information, please visit our website.

Why subscribe?

- Spend less time learning and more time coding with practical eBooks and Videos from over 4,000 industry professionals

- Improve your learning with Skill Plans built especially for you

- Get a free eBook or video every month

- Fully searchable for easy access to vital information

- Copy and paste, print, and bookmark content

Did you know that Packt offers eBook versions of every book published, with PDF and ePub files available? You can upgrade to the eBook version at packtpub.com and as a print book customer, you are entitled to a discount on the eBook copy. Get in touch with us at customercare@packtpub.com for more details.

At www.packtpub.com, you can also read a collection of free technical articles, sign up for a range of free newsletters, and receive exclusive discounts and offers on Packt books and eBooks.

Share Your Thoughts

Now you've finished *System Design Guide for Software Professionals*, we'd love to hear your thoughts! Scan the QR code below to go straight to the Amazon review page for this book and share your feedback or leave a review on the site that you purchased it from.

`https://packt.link/r/1-805-12499-4`

Your review is important to us and the tech community and will help us make sure we're delivering excellent quality content.

Download a free PDF copy of this book

Thanks for purchasing this book!

Do you like to read on the go but are unable to carry your print books everywhere?

Is your eBook purchase not compatible with the device of your choice?

Don't worry, now with every Packt book you get a DRM-free PDF version of that book at no cost.

Read anywhere, any place, on any device. Search, copy, and paste code from your favorite technical books directly into your application.

The perks don't stop there, you can get exclusive access to discounts, newsletters, and great free content in your inbox daily

Follow these simple steps to get the benefits:

1. Scan the QR code or visit the link below

https://packt.link/free-ebook/9781805124993

2. Submit your proof of purchase

3. That's it! We'll send your free PDF and other benefits to your email directly

www.ingramcontent.com/pod-product-compliance
Lightning Source LLC
LaVergne TN
LVHW081513050326
832903LV00025B/1473